WAR AT THE SPEED OF LIGHT

WAR AT THE SPEED OF LIGHT

DIRECTED-ENERGY WEAPONS AND THE FUTURE OF TWENTY-FIRST-CENTURY WARFARE

LOUIS A. DEL MONTE

Potomac Books

AN IMPRINT OF THE UNIVERSITY OF NEBRASKA PRESS

Library of Congress Cataloging-in-Publication Data
Names: Del Monte, Louis A., author.
Title: War at the speed of light: directed-energy weapons and
the future of Twenty-First-Century warfare / Louis A. Del
Monte.
Description: Lincoln: Potomac Books, an imprint of the
University of Nebraska Press, [2021] | Includes bibliographical
references and index.
Identifiers: LCCN 2020017971
ISBN 9781640123304 (hardback)
ISBN 9781640124356 (epub)
ISBN 9781640124363 (mobi)
ISBN 9781640124370 (pdf)
Subjects: LCSH: Directed-energy weapons. | Military art
and science—History—21st century. | Military art and
science—Forecasting.
Classification: LCC UG486.5 .D45 2021 | DDC 355.8/246—dc23
LC record available at https://lccn.loc.gov/2020017971

Set in Arno Pro by Laura Buis.

After half a century of marriage, it is with admiration for her intelligence, honesty, and wisdom that I dedicate this book to my wife and love of my life, Diane E. Del Monte

In loving memory of my dear friend Nick McGuinness (1939–2020), a remarkable man who, by example, taught us how to live a meaningful life and face death with courage and dignity

CONTENTS

List of Illustrations ix

List of Tables. x

Acknowledgments xi

Introduction 1

Part 1. The Game of Cat and Mouse

1. Tilting the Balance of Terror 13

2. The Quest for Global Dominance Using
   Conventional Weapons 27

3. The Coming Fourth U.S. Offset Strategy 47

Part 2. Directed-Energy Weapons

4. Laser Weapons. 65

5. Microwave Weapons 89

6. EMP Weapons 111

7. Cyberspace Weapons. 129

Part 3. "Shields Up, Mr. Sulu"

8. Directed-Energy Countermeasures 153

9. Force Fields. 167

Part 4. The Coming New Reality

10. Autonomous Directed-Energy Weapons 179

11. The Full-Scale Weaponization of Space. 193

12. Not Gambling with the Fate of Humanity 213

Appendix A: U.S. and Chinese Defense Budgets Adjusted
for Purchasing Power Parity and Labor Costs. 223

Appendix B: The Design and Operation of a Laser 225

Appendix C: Radiation-Hardened Electronics and System
Shielding Resources 227

Appendix D: Articles Describing the Operation of
Nonnuclear EMP Devices. 229

Notes 231

Index. 259

ILLUSTRATIONS

1. U.S. Navy laser weapon system 45
2. Russia's truck-mounted laser weapon 87
3. The x-37b Orbital Test Vehicle 208

TABLES

1. Comparison between high-power laser and high-power microwave anti-satellite weapons 205

ACKNOWLEDGMENTS

With a deep appreciation I acknowledge the many contributions of my wife, Diane E. Del Monte, who used her liberal education as well as her talents as an artist and art teacher to edit and help form this book. She is the bedrock of our family and has, by example, taught us the meaning of unconditional love. In art and life, Di's compass always pointed our way to true north. She has continually put our family's needs beyond those of her own. Through good times and challenging times, Di's unique out-of-the-box thinking has helped all those around her persevere. Her compassion and generosity toward others, especially those in need, is exemplary. Di is and will remain my soul mate.

I also want to recognize the many contributions of Nick McGuinness, whose friendship I cherish. Nick is one of the best-read and most knowledgeable people I know, and he brought his wealth of education to help refine this book. His chapter-by-chapter comments and edits were invaluable. I will be forever in his debt.

If it were not for the skill of my agent, Jill Marsal, this book would not have found a publisher. Jill is a founding partner of the Marsal Lyon Literary Agency, whose goal is to build long-term relationships with its authors and publishing partners. This book is the third Jill has helped bring to market, proving beyond doubt that the agency is achieving its goal. I am grateful for her guidance and representation.

Last, I would like to thank the team at Potomac Books, the publisher that brought this book to the public, and Vicki Chamlee, who edited the manuscript and improved its readability.

WAR AT THE SPEED OF LIGHT

Introduction

A s I began my graduate work in September 1966, a new science fiction television program, *Star Trek*, debuted. I was majoring in physics, and I loved science fiction. Gene Roddenberry was the show's creator, and fans now refer to it as the "Original Series." Its simple premise was to follow the voyages of the starship USS *Enterprise* on its five-year mission, which was "to explore strange new worlds, to seek out new life and new civilizations, to boldly go where no man has gone before."

The *Star Trek* series became immensely popular. The formidable weapons of the USS *Enterprise*, including phasers, photon torpedoes, and an invisible energy shield that protected the vessel during a battle, awed me. Although Roddenberry never used the term "directed-energy weapons," that is what they were. It did not cross my mind that someday I would work on and write about similar weapons.

War at the Speed of Light describes the ever-increasing and revolutionary role of directed-energy weapons in warfare, including laser, microwave, electromagnetic pulse (EMP), and cyberspace weapons. Additionally, and most important, this book delineates the threat that directed-energy weapons pose to disrupting the doctrine of mutually assured destruction (MAD), which has kept the major powers of the world from engaging in a nuclear war since World War II.

The Changing Nature of Warfare

The nature of warfare is changing in three fundamental ways.

1

Artificial intelligence (AI) is quickly replacing human intelligence both in initiating the release of weapons as well as in their target acquisition. Initially, we termed such weapons as "smart," which connoted they emulated human intelligence, albeit in the narrow parameters of their early deployment. As the level of AI increased, we witnessed the emergence of semiautonomous and autonomous weapons. Semiautonomous weapons require a human to release them and to exercise final judgment regarding the taking of human life. Autonomous weapons act on their own, meaning they can self-initiate and take human life without human intervention. Semiautonomous and autonomous capabilities elevated the weapons' performance to the speed of computer technology, outpacing human reaction time on the twenty-first-century battlefield.

While the ethics of autonomous weapons is subject to significant debate, countries such as the United States, the People's Republic of China (hereafter China), and Russia, among others, are actively engaged in developing and deploying weapons with ever-increasing AI. The reason for this is simple: the use of AI weapons has already proved highly effective on the battlefield. One vivid example is the U.S. military's use of smart bombs in the First Gulf War against Iraq. Millions of people watched CNN's live newscasts from the battlefield and footage from cameras on board U.S. bombers demonstrating the precision of the aerial laser-guided bombs. In fact, AI has proved so critical in making weapons more effective the U.S. Department of Defense (DOD) adopted it as a key pillar in its Third U.S. Offset Strategy (i.e., in addition to strengthening conventional forces for deterrence, it incorporated next-generation technologies to change an unattractive competitive position to a more advantageous one).

To establish the Third U.S. Offset Strategy, Defense Secretary Chuck Hagel issued a memorandum on November 15, 2014, delineating the military's quest for a combination of new technologies to maintain America's military supremacy. Top priorities, according to Hagel, are "robotics, autonomous systems, miniaturization, big data, and advanced manufacturing, including 3-D printing."[1]

Given Hagel's priorities, the Third U.S. Offset Strategy may appear complicated and all-inclusive. However, on October 31, 2016, during the Center for Strategic and International Studies' event titled "Assessing the Third Offset Strategy," Deputy Defense Secretary Bob Work gave us valuable insight. He told attendees, "The third offset's initial vector is to exploit all the advances in artificial intelligence and autonomy and insert them into DOD's battle networks to achieve a step increase in performance that the department believes will strengthen conventional deterrence."[2]

From this, we can see the Third U.S. Offset Strategy relies heavily on AI, which enables autonomous systems. AI is also allowing the military to take the human out of the weapon, thus enabling the weapon to become physically smaller. For example, consider the U.S. Air Force's stealth bomber. If the United States built a stealth bomber with AI equal to human intelligence, it would be smaller because the life support elements and armor protection a human pilot requires could be removed. Such an aircraft would align well with the Third U.S. Offset Strategy's goal of miniaturization.

Today's military strategists plan to have scores of small, expendable, autonomous weapons. While historically the military was on a quest for the biggest and most survivable weapons, with the advent of AI, robotics, and miniaturization, it is now able to borrow an effective tactic from nature—namely, "swarming." In simple terms, the U.S. military plans to use swarms of cost-effective autonomous weapons that will overwhelm an adversary's defensive capabilities. In this context, I use the phrase "cost-effective autonomous weapons" to imply low-cost and potentially expendable autonomous weapons.

Due to the quickening pace of war enabled by AI and swarming tactics, the U.S. military is turning to a new class of weapons for defense—namely, directed-energy weapons. This new military thrust is forging the second fundamental change in warfare.

Projectile weapons, such as missiles, inflict damage either by their kinetic energy—a function of their mass and velocity—or by their explosive payload. Directed-energy weapons, such as lasers, inflict damage by focusing electromagnetic energy (i.e., light) on their intended target. The critical distinction is that while all weapons ultimately rely on energy to disable or destroy their intended targets, directed-energy weapons project devastation at the speed of light. Even a hypersonic missile traveling at five times the speed of sound (about a mile per second) would appear to be standing still compared to a laser pulse traveling to its intended target at the speed of light (approximately 186,000 miles per second).

Potential adversaries of the United States, such as China and Russia, are developing and deploying supersonic (i.e., faster than the speed of sound) and hypersonic missiles as a means to destroy U.S. aircraft, drones, missiles, aircraft carriers, and space-based assets such as the Global Positioning System (GPS) and communications satellites. To counter this threat, the United States is developing and deploying laser weapons; however, the development of these weapons is in its infancy. For example, in December 2014, the U.S. Navy installed the first-ever laser weapon on the USS *Ponce*. In field testing, the navy reported the laser system worked perfectly against low-end asymmetric threats, such as small unmanned aerial vehicles. Following the field tests, the navy authorized the commander of the *Ponce* to use the system as a defensive weapon. This is just the beginning. The U.S. Navy's strategy is to develop higher-energy laser systems with the capability to destroy an adversary's "carrier killer" missiles as well as other asymmetric threats such as hypersonic missiles. In January 2018 the navy contracted Lockheed Martin to deliver two high-energy lasers with integrated optical-dazzler and surveillance (HELIOS) systems by 2021. The navy intends to deploy one on the *Arleigh Burke*–class destroyer USS *Preble*. The other will be land based at the White Sands Missile Range in New Mexico for testing. In the 2020s, the U.S. military plans to usher in the widespread use of laser weapons on land, at sea, and in the air and space. One can reasonably assume these new lasers will continue the U.S. military's thrust to develop and deploy laser weapon systems capable of destroying an adversary's supersonic, hypersonic, and intercontinental ballistic missiles (ICBMS); drone swarms; and space assets.

In addition to lasers, the U.S. military is pursuing a full spectrum of directed-energy weapons, including microwave, EMP, and cyberspace weapons. At this point, you may wonder, why make the rapid move to directed-energy weapons?

In the quest for global dominance, a vicious game of cat and mouse is taking place between the United States, China, and Russia. Both China and Russia are keenly aware that the United States holds a leadership position in aircraft carrier groups, stealth aircraft, and space assets. Unable to match the U.S. military's dominance in these weapon systems, China and Russia are focusing on asymmetrical warfare with weapons that attack areas of the U.S. military's weakness.

For China, asymmetric warfare represents a tactic with ancient roots dating back to Sun Tzu's *The Art of War*. As Tzu wrote, "If your enemy is secure at all points, be prepared for him. If he is in superior strength, evade him. If your opponent is temperamental, seek to irritate him. Pretend to be weak, that he may grow arrogant. If he is taking his ease, give him no rest. If his forces are united, separate them. If sovereign and subject are in accord, put division between them. Attack him where he is unprepared, appear where you are not expected."[3] So in war, the way to succeed is to avoid what is strong and strike at what is weak.

Both China and Russia are taking asymmetric warfare to the next level. For example, China announced in April 2018 the activation of a new brigade of Beijing's most advanced intermediate-range ballistic missile, the Dong Feng-26. According to *Missile Threat*, an online source for information and analysis about ballistic and cruise missiles, the Dong Feng-26 "is China's first conventionally-armed ballistic missile capable of striking Guam. It has a range of 3,000–4,000 km [about 1,800–2,400 miles], capable of ranging most U.S. military bases in the eastern Pacific Ocean. The missile can be armed with a conventional or nuclear warhead, and an antiship variant may also be in development."[4]

Russia likewise has directed efforts to offset the U.S. military's capability in stealth aircraft, deploying in 2007 the s-400 surface-to-air defense missile system. The *Economist* described it as "one of the best air-defense systems currently made."[5]

As the most capable potential adversaries of the United States deploy missile defenses that could threaten its advanced weapons systems, such as its *Ford*-class aircraft carriers and b-2 stealth bombers, the U.S. military is developing countermeasures. Current countermeasures rely on antiballistic missile defense systems, such as the Terminal High Altitude Area Defense (THAAD). These countermeasures primarily use missiles to destroy missiles, which is akin to using bullets to stop bullets.

Unfortunately, these countermeasures do not cover the complete threat spectrum. For example, THAAD is only effective against short-, medium-, and intermediate-range ballistic missiles, not against ICBMs. Also, the countermeasures can be an expensive deterrent. For example, in 2017, a U.S. ally used a Patriot missile, priced at about $3 million, to shoot down a small enemy quadcopter drone that is available on Amazon for about $200. Of course, the quadcopter drone had no chance against the Patriot, a radar-targeted missile more commonly used to shoot down enemy aircraft and ballistic missiles. The military terms this usage "overkill." In theory, an enemy could order more of the cheap quadcopter drones online until it exhausts the U.S. and allied militaries' stock of Patriot missiles.

Given the expense of using missiles to counter enemy missiles and drones, along with their ineffectiveness across the entire threat spectrum, the U.S. military is turning to laser and other directed-energy weapons. While the price tag for hypersonic missiles continues to soar, approaching $600 million per missile, the cost per laser pulse continues to drop, approaching about a dollar per shot. In addition, as we explore in this book, the U.S. military feels that directed-energy weapons will be effective against the entire threat spectrum, from ICBMs to drone swarms.

One subtle point that may not be apparent is that missile defenses, drones, drone swarms, and personnel rely on electromagnetic energy, such as radio signals and radar, for guidance and communications. This reliance on electromagnetic energy brings us to the third fundamental change in warfare.

Cyberspace Is Officially a Battlefield

In June 2016, the North Atlantic Treaty Organization (NATO) declared cyberspace (computer networks and the internet) as an "operational domain," a battlefield as real as air, sea, land, and space. This declaration recognizes cyber warfare and electronic warfare as two crucial new elements of warfare.

To succeed in this new battle space, the U.S. military must be equipped with capabilities to defend or attack information networks in cyberspace (i.e., cyber warfare) and to control access to the electromagnetic spectrum (i.e., electronic warfare). As a result, the U.S. military is integrating cyber and electronic warfare to achieve an effective defense and offense in this new battle space.

Cyber warfare typically involves operations that disrupt, exploit, or cripple adversaries through information systems and the internet via the use of computer code and computer applications. It often includes launching cyber weapons wirelessly, or transmitting cyber weapons as electromagnetic radiation, similar to radio waves, traveling at the speed of light.

During the first decade of the twenty-first century, cyber warfare was the stuff of theoretical scenarios by security professionals. However, as this book demonstrates, now hackers can cause just as much damage as traditional military attacks.

Electronic warfare is military action involving the use of directed energy to control the electromagnetic spectrum—such as radar, radio transmissions, and laser beams—either to deceive or attack an enemy or to protect friendly systems from similar actions. The goal, according to the Department of Defense, is to use directed-energy weapons to disrupt an electromagnetic field, resulting in jamming and deceiving information managed by computerized systems or electronic platforms such as surveillance or telecommunication satellites. With high power, these weapons can also burn out the electric circuitry of an adversary's weapon, resulting in the destruction or interference of its function.

The Bottom Line

The pace of warfare is accelerating. In fact, according to the Brookings Institution, a nonprofit public policy organization, "So fast will be this process [command and control decision-making], especially if coupled to automatic decisions to launch artificially intelligent autonomous weapons systems capable of lethal outcomes, that a new term has been coined specifically to embrace the speed at which war will be waged: hyperwar."[6]

The term "hyperwar" adequately describes the quickening pace of warfare resulting from the inclusion of AI into the command, control, decision-making, and weapons of war. However, to my mind, it fails to capture the speed of conflict associated with directed-energy weapons. To be all inclusive, I would like to suggest the term "c-war." In Albert Einstein's famous mass-energy equivalent equation, $E = mc^2$, the letter c is used to denote the speed of light in a vacuum. (For completeness, E means energy and m, mass.) Surprisingly, the speed of light in the earth's atmosphere is almost equal to its velocity in a vacuum. On this basis, I believe c-war more fully captures the new pace of warfare.

Unfortunately, c-war, or war at the speed of light, may remove human judgment from the realm of war altogether, and that could have catastrophic ramifications. If you think this is farfetched, consider the following Cold War account, where new technology almost plunged the world into nuclear war:

Lt. Col. Stanislav Petrov settled into the commander's chair in a secret bunker outside Moscow. His job that night was simple: Monitor the computers that were sifting through satellite data, watching the United States for any sign of a missile launch. It was just after midnight, Sept. 26, 1983.

A siren clanged off the bunker walls. A single word flashed on the screen in front of him.

"Launch."

Petrov's computer screen now showed five missiles rocketing toward the Soviet Union. Sirens wailed. Petrov held the phone to the duty officer in one hand, an intercom to the computer room in the other. The technicians there were telling him they could not find the missiles on their radar screens or telescopes.

It didn't make any sense. Why would the United States start a nuclear war with only five missiles? Petrov raised the phone and said again:

"False alarm."

For a few terrifying moments, Stanislav Petrov stood at the precipice of nuclear war. By mid-1983, the Soviet Union was convinced that the United States was preparing a nuclear attack. The computer system flashing red in front of him was its insurance policy, an effort to make sure that if the United States struck, the Soviet Union would have time to strike back.

But on that night, it had misread sunlight glinting off cloud tops.

"False alarm." The duty officer didn't ask for an explanation. He relayed Petrov's message up the chain of command.[7]

The world owes Lt. Col. Stanislav Petrov an incalculable debt. His judgment spared the world a nuclear holocaust. Now ask yourself this simple question: if those systems Petrov monitored were autonomous (i.e., artificially intelligent), would they have initiated World War III? I believe it is a profound question, and we could make persuasive arguments on either side. However, would you want to leave the fate of the world to an artificially intelligent system?

I have devoted a significant portion of my career to developing AI for military applications. My experience leads me to conclude today's technology cannot replicate human judgment; therefore, I think if an AI system had replaced Petrov, it may have initiated World War III. I also believe U.S. military planners are acutely aware of this possibility and are taking steps to defend the United States against such a mishap. Such actions could disrupt the doctrine of MAD, which prevents nuclear war via the threat of mutually assured destruction, or what some term as "the balance of terror." If any country were able to disrupt the doctrine of MAD, then it would tilt the balance of terror. How this might happen is one page away.

ONE

THE GAME OF CAT AND MOUSE

1

Tilting the Balance of Terror

The balance of terror has replaced the balance of power.

—LESTER PEARSON

The devastation of war is always about energy. This statement is true historically as well as to this day. For example, most of the massive destruction during World War II resulted from dropping conventional bombs on an adversary. To understand the role energy plays in this type of devastation, consider the Japanese attack on Pearl Harbor. On December 7, 1941, Imperial Japan launched 353 bombers and torpedo bombers in two waves from six aircraft carriers.[1] Their bombs and torpedoes incorporated trinitroanisole, a chemical compound.[2] The vast devastation caused by unleashing the energy in trinitroanisole's chemical compound resulted in the sinking of twelve ships and damaging nine others. The attacks also destroyed 160 aircraft and damaged another 150, and more than 2,300 Americans lost their lives.[3]

A near-perfect example of energy's devastation is the atomic bombings of Hiroshima and Nagasaki on August 6 and August 9, 1945, respectively. These bombs were different from all those that preceded them. They derived their destructive force from nuclear fission, or the splitting of atoms. In simple terms, energy holds an atom together, and when a fast-moving subatomic particle causes the atom to split into its subatomic particles, it releases the energy binding the atom together. We know from Einstein's famous mass-energy equivalent formula, $E = mc^2$, that even a small amount of mass (m) converted to energy (E) yields an enormous amount of energy. The reason for this is that mass is multiplied by the speed of light (c) squared (i.e., times itself). The velocity of light is approximately equal to 186,000 miles per second. Doing the math yields an enormous amount of energy from a relatively small amount of mass. Examining the atomic bombs demonstrates this point. Each used fissionable material measuring less than two hundred pounds, yet they unleashed the devastation of fifteen thousand to twenty thousand tons of TNT.

I know it is unusual to think about destruction as being related to energy, but that is a fact of war. From the first caveman who used a rock to kill an adversary to a sniper's firing a bullet, it all has to do with energy. In the case of the rock and the bullet, their kinetic energy (a function of their mass and velocity) inflicts wounds. Think of any weapon, except biological and chemical weapons, from the earliest times to the present, and you face one inescapable conclusion: it relies on some form of energy to carry out its mission.

If you are a fan of the science fiction series *Star Trek*, you are aware that the starship *Enterprise* and its crew did not use anything that resembled conventional weapons, such as guns, or nuclear weapons. Also, the *Enterprise* did not have traditional armor plating. In *Star Trek*, we see the crew using handheld phasers that could be set to kill or to stun. The phasers, set to kill, are a fictional extrapolation of real-life lasers. When set to stun, the phasers are comparable to real-life microwave weapons that have a stunning effect.[4] In place of missiles, the *Enterprise* fired photon torpedoes, which are similar to the missiles current military warplanes and warships fire, except the torpedo's warhead was not a conventional or nuclear explosive. The photon torpedo's warhead consisted of antimatter, which has the destructive property of annihilating matter, or converting it to energy. Last, in place of armor plating to shield the ship, the *Enterprise* used a fictional force field, which is similar to the real-life active protection systems deployed to protect U.S. military vehicles.[5] In essence, Gene Roddenberry's *Star Trek* exposed viewers to directed-energy weapons.

Although *Star Trek* is purely fictional, directed-energy weapons are becoming a reality. In their genius, Roddenberry and the writers of *Star Trek* combined what the National Aeronautics and Space Administration (NASA) termed an "entertaining combination of real science, imaginary science gathered from lots of earlier stories, and stuff the writers make up week-by-week to give each new episode novelty."[6] When the first episode of the original series aired on September 6, 1966, on CTV in Canada and September 8, 1966, on NBC in the United States, the weapons of *Star Trek* were complete fiction. However, like all good science fiction, *Star Trek*'s weapons were an extrapolation of things to come. Today, many of *Star Trek*'s weapons are either deployed or in development, with life imitating art.

If you find the prospect of real-life, *Star Trek*–type directed-energy weapons scary, then be prepared to be scared. The United States is leading the world in developing and deploying directed-energy weapons. As the U.S. military deploys these new weapons, such as lasers that can destroy nuclear-tipped ballistic missiles in flight, expect doctrines such as that of mutually assured destruction to crumble and the balance of terror to tilt.

To understand how severe a directed-energy weapon attack could be, consider the following possible scenario.

Scenario: Small-Town USA Somewhere on the East Coast

John found waking up in the morning difficult, and this morning was no exception. He was surprised to see his old-fashioned alarm clock did not go off, and the second hand was standing still at precisely five seconds after 4:00 a.m.

Damn, he thought. *I'm going to be late, and my alarm clock is broken.*

He got up, sat on the edge of the bed, and reached to turn on the light on his nightstand. The light failed to work.

His mind raced. *What the hell, the light is also broken? Maybe the power is out.*

He reached for his Seiko 5 wristwatch, an automatic mechanical watch that uses a wound spring as a power source. In the dimly lit room, the glow of the watch's hands revealed it was 8:36 a.m.

He moved through the house, checking the light switches and the television. Nothing worked. Even his old, handheld portable radio only provided static. At this point, he knew something was wrong, but he had no idea what.

He opened the drapes to let the light in and got dressed. To his surprise, he saw several of his neighbors talking in the street. He decided to join them.

Approaching, he thought it odd that one neighbor was holding an AR-15, commonly known as an assault rifle.

"Is everything all right? Is your power out too?" John asked as he joined them. He made no mention of the AR-15.

"Everything is out. Nothing is working. My car won't even start," replied one of the men.

"What's going on?" he asked, hoping they had some answers.

"An EMP has hit us," said the neighbor holding the AR-15.

"A what?"

"An electromagnetic pulse . . . EMP."

"What's that?"

"I'm not a physicist, but I know a nuclear device detonating in the atmosphere can cause it. The resulting radiation burns out all circuits . . ."

"How do you know about this EMP?"

"I had my transistor radio in my gun safe, shielded, and the government's emergency broadcasting network has been going non-stop about this EMP attack."

The word "attack" caught John's attention. "Are we under attack?"

"That's my take."

John could feel his blood pressure rising. "What do they say we should we do?"

"They recommend staying indoors and waiting for further instructions, but that's just bull. I've filled my bathtub with water and walked to the 7-Eleven and bought everything I could carry."

John was confused and asked, "Why did you fill your bathtub with water?"

"The water tower is going to run dry, and they have no way to pump more water into it. It's best to have some on hand for drinking and cooking."

That made sense to John, but he wondered about the rifle. "What's the rifle for?"

"The 7-Eleven was a mess. People were climbing over people and grabbing everything in sight, even out of other people's arms. Some didn't even pay." Then his neighbor brought his AR-15 to chest height. "I think we're on our own now . . ."

Somewhere in the deep recesses of his mind, John remembered reading an article about a potential EMP attack. The one thing that remained seared on his brain is that 90 percent of the population would be likely to die in the first year if the EMP attack was nationwide.

"Did the emergency alert say anything about it being a nationwide attack?" John asked.

"No, it just kept repeating to stay inside and wait for further instructions."

More people were coming out of their houses, asking the same questions, obviously as similarly confused as John was.

Suddenly John could feel fear paralyzing his mind. He was numb, almost glued in place. He now realized that there was no way to dial 911 for help or if there were even emergency personnel or cops still on the job. Even worse, he worried that the United States was at war. His mood deepened to a sense of dread.

Is this how it ends? he wondered with tears in his eyes.

While the scenario is fictional, it is potentially possible. Although chapter 6 is devoted to EMP weapons, I think it is instructive here to explain briefly how such an event could happen.

One possible way to wage an EMP attack is for a country to detonate a nuclear device about three hundred miles above a particular landmass and produce an abrupt pulse of high-energy electromagnetic radiation and intense gamma radiation. The electromagnetic radiation will result in damaging current and voltage surges in the area's electrical and electronic systems. The intense gamma radiation ionizes the atmosphere, which also causes an EMP as the atoms of air first lose their electrons and then regain them. The mechanism of destroying electrical and electronic systems is the same in both cases. The nuclear detonation results in rapidly changing electric fields, which in turn cause rapidly changing high electrical surges that lead to further damage. Indeed, the damage can even extend to the electrical grid.[7]

In the preceding scenario, a high-altitude nuclear detonation resulted in damage to electrical and electronic systems as well as the power grid. You may wonder, Why did the transistor radio work after the attack? In simple terms, the transistor radio was in a metal gun safe, which acted as a Faraday cage—that is, an enclosed, grounded metal screen—and shielded it from the EMP attack. The Seiko 5 wristwatch worked because it was a mechanical watch, not quartz. A quartz watch would have burned out. We explain further in chapter 6, but for now, recognize that anything electric or electronic is vulnerable during an EMP attack unless shielded by an enclosed metal case, such as a gun safe or a bare metal garbage can.

The Balance of Terror

Many use the phrase "balance of terror" to describe the nuclear arms race between the United States and the Union of Soviet Socialist Republics (USSR) during the Cold War. Both countries possessed sufficient nuclear weapons not only to annihilate each other but also to severely damage all of human civilization. Lester Pearson coined the phrase in June 1955 at the tenth anniversary of the signing of the United Nations (UN) Charter when he stated, "The balance of terror has replaced the balance of power."

Some political scientists use the phrase "balance of terror" to differentiate the world situation that followed World War II. Before World War II, nations prevented war by maintaining a relative balance of their ability to wage war against each other. The phrase "balance of power" characterized this tenuous peace. The atomic bomb created a new world reality. For the first time in history, the two superpowers of the Soviet Union and the United States could destroy each other and every living thing. Pearson, in coining the phrase, captured the feeling of terror this situation evoked throughout the world.

This uneasy tenuous peace is where we find ourselves as of this writing. Fortunately, no nation has detonated nuclear weapons in conflict since the atomic bombing of Nagasaki in 1945. However, before we congratulate ourselves, we must recognize that nuclear weapons did not make the world a peaceful place. In fact, according to the Peace Pledge Union, a nongovernmental organization that promotes pacifism, "Since the end of the Second World War in 1945 there have been some 250 major wars in which over 50 million people have been killed, tens of millions made homeless, and countless millions injured and bereaved."[8]

The existence of nuclear weapons did make it impossible for capable potential adversaries such as Russia, China, and the United States to engage in a full-scale world war. Nonetheless, they are still engaged in an all-out arms race to have the most effective conventional (i.e., nonnuclear) weapons, and the world faces a New Cold War, sometimes referred to as Cold War II.

Cold War II

Cold War II is an ongoing state of political and military tension between opposing geopolitical power blocs. Based on current military capabilities, Russia and China reportedly are leading one bloc. The United States, European Union, and NATO are leading the opposing bloc. Similar to the original Cold War, we see standoff and proxy wars between the different blocs. For example, the civil war to overthrow Bashar al-Assad's regime in Syria has been widely described as a series of overlapping proxy wars among world powers, primarily between the United States and Russia prior to the U.S. withdrawal from the conflict.[9]

As noted, despite the Cold War II and the numerous proxy wars, no nuclear power has used nuclear weapons in a conflict, as the concern is that even the limited use of nuclear weapons could unleash a full-scale nuclear exchange. This restraint on the part of capable nuclear adversaries, such as Russia, China, and the United States, is not the result of a treaty; instead, it relies on the unwritten doctrine of mutually assured destruction. In reality, MAD is more of a military strategy that ensures the full-scale use of nuclear weapons by two or more opposing sides would cause the complete annihilation of both the attacker(s) and the defender(s).

The End of MAD

Evidence shows the military strategy of MAD is changing. For example, the administration of U.S. president George W. Bush withdrew from the Anti-Ballistic Missile (ABM) Treaty in June 2002.[10] This arms control treaty between the United States and the Soviet Union put limitations on the antiballistic missile systems used in defending areas against ballistic missile–delivered nuclear weapons. It was, in principle, a method to ensure MAD remained a viable defense strategy. In withdrawing from the ABM Treaty, the Bush administration claimed that the United States' limited national missile defense system, which it proposed to build, would prevent nuclear blackmail by a state with limited nuclear capability—for example, North Korea—and would not alter the nuclear posture between Russia and the United States.

According to *International Security*, a peer-reviewed academic journal in the field of international and national security, the doctrine of MAD may have run its course:

> The age of MAD, however, is waning. Today the United States stands on the verge of attaining nuclear primacy vis-à-vis its plausible great power adversaries. For the first time in decades, it could conceivably disarm the long-range nuclear arsenals of Russia or China with a nuclear first strike. A preemptive strike on an alerted Russian arsenal would still likely fail, but a surprise attack at peacetime alert levels would have a reasonable chance of success. Furthermore, the Chinese nuclear force is so vulnerable that it could be destroyed

even if it were alerted during a crisis. To the extent that great power peace stems from the pacifying effects of nuclear weapons, it currently rests on a shaky foundation.[11]

What is causing this apparent change in military strategy regarding the doctrine of MAD? In a word, technology. Military strategy evolves as technology evolves. Six significant technology evolutions are causing military strategists to think beyond MAD.[12]

1. Cyber Warfare: As discussed in the introduction, the cyber domain is becoming a battlefield. Leading its militarization are the United States, China, and Russia. The militarization of cyberspace could pose a severe threat to MAD. For example, if capable potential adversaries such as China or Russia can insert malicious code into U.S. early warning and nuclear command-and-control satellites, it could delay and even prevent the United States from being able to retaliate before its destruction. Alternately, potential adversaries could use directed electromagnetic energy to jam these satellites. In either case, it would render the United States' capability of launching a nuclear counterstrike before its destruction questionable. However, unlike China and Russia, the U.S. military maintains half of its nuclear capability in its stealthy nuclear submarines scattered secretly around the globe.[13] This capability would probably enable a counterstrike, but as becomes evident in later chapters, human life on Earth might cease to exist. Thus, the nuclear deck appears stacked in favor of the United States, which may also use cyber weapons to disrupt the nuclear command and control of its potential adversaries, further rendering the doctrine of MAD useless.

2. Anti-satellite Missiles: According to a Joint Chiefs of Staff J-2 (Intelligence Directorate) report, China and Russia are developing anti-satellite missiles that will soon be capable of damaging or destroying U.S. satellites in low-earth orbit.[14] This capability could destroy the early warning and nuclear command-and-control satellites of the United States. Unlike a cyberattack, such destruction would be a visible act of war that could unleash a full nuclear counterstrike by the United States. An adversary using this tactic would

be betting that the United States would be unable to defend itself "unplugged." I think that is a foolish bet. U.S. counterstrike capabilities are not solely dependent on space assets. For example, intercontinental ballistic missiles are self-guided munitions.[15] Still, anti-satellite missiles are concerning. The United States, more than any other nation, depends on its space assets to fight wars conventionally. The loss of such assets would make the doctrine of MAD problematic by forcing the United States to choose nuclear war or defeat. (See also chapter 11.)

3. Directed-Energy Weapons: As discussed in the introduction, the United States and its potential adversaries are developing directed-energy weapons, such as laser and microwave weapons. The United States appears to be a clear leader, having already deployed a laser weapon system in 2014 on USS *Ponce* that is capable of destroying small boats and cruise missiles. Although not stated by the Department of Defense, I also think the U.S. military likely is considering or may have even deployed a laser weapon in space. The United States is a signatory of the Treaty on Principles Governing the Activities of States in the Exploration and Use of Outer Space, including the Moon and Other Celestial Bodies, or the Outer Space Treaty, which does not prohibit deploying conventional weapons in orbit but does ban deploying weapons of mass destruction in orbit. Directed-energy weapons are conventional weapons, not weapons of mass destruction.[16]

For this reason, the United States refused to sign the proposed 2008 Chinese-Russian Treaty on the Prevention of the Placement of Weapons in Outer Space and of the Threat or Use of Force against Outer Space Objects (generally referred to as PPWT) and its later variants.[17] As is likely clear to U.S. military strategists, directed-energy weapons would be ideal for shooting large numbers of nuclear missiles out of the sky, thereby defeating swarming, first-strike, and anti-satellite weapons strategies.[18] However, this tactic would require both the directed-energy weapons to mature and the United States to deploy them in space. It represents a perfect example of the military cat-and-mouse game among nations: China and Russia are developing anti-satellite weapons as an asymmetrical strategy to destroy the ability of the United States to retaliate against a first-strike nuclear attack, while the United States is developing a countermeasure strategy—namely, the use of directed-energy weapons—to destroy any missiles that threaten its space assets as well as those of its NATO allies.

4. Rail Guns: While the name may sound as if it is some gun carried by a railroad car, à la World War II–type weapons, nothing could be further from reality. A rail gun uses an electromagnetic field to launch a projectile at five to seven times the speed of sound, making it significantly faster than a conventional, explosive-powered military gun's projectile.[19] Even though the rail gun's projectile typically carries no explosives, it can achieve more destruction than a conventional explosive-powered military gun launching a similar-size projectile. The destructive force of the rail gun is a function of the kinetic energy of its projectile (i.e., the mass times the square of the velocity). Since a rail gun launches a projectile at hypersonic speeds, its trajectory is flatter than that of a conventional, explosive-powered military gun. Also, it can hit a target a hundred miles away. In the rail gun, the U.S. military can launch a hypervelocity projectile that can penetrate over six inches of steel and three reinforced concrete walls. While the navy already has missiles that perform the same feats, they cost millions of dollars each. The *Diplomat*, an online international news magazine in the Asia-Pacific region, made this point: "Traditional missile projectile systems on U.S. ships can range in costs from $500,000 to $1.5 million. The railgun projectile, weighing at roughly 23 lbs, costs $25,000, and can be fired at speeds of Mach 7 or 5,000 miles per hour."[20]

The U.S. Navy is eyeing the rail gun for its latest destroyer, the USS *Zumwalt*. According to Rear Adm. Matt Klunder, the rail gun "will give our adversaries a huge moment of pause to go: 'Do I even want to go engage a naval ship?' Because you are going to lose. You could throw anything at us, frankly, and the fact that we now can shoot a number of these rounds at a very affordable cost, it's my opinion that they don't win."[21] Klunder added, "There's not a thing in the sky that's going to survive against that."[22]

In general, U.S. military strategists see the rail gun as a way to shoot large numbers of nuclear missiles out of the sky.[23] Unfortunately, other militaries see the rail gun in much the same way. For example, as of January 2019, a Chinese naval warship at sea is reportedly carrying what appears to be an electromagnetic rail gun. If operational, it would put the Chinese ahead of the United States on deploying rail guns.[24]

5. Hypersonic Missiles: The United States, Russia, and China are investing in an entirely new class of weaponry—namely, highly precise, long-range, hypersonic missiles. Hypersonic missiles are missiles that can fly at more than five times the speed of sound, travel lower in the atmosphere than current nuclear-tipped ballistic missiles, and maneuver during flight. Hypersonic missiles pose a threat to the doctrine of MAD in that they will allow a nation to strike an adversary in a matter of minutes. Also, their low trajectory will enable them to travel farther and more stealthily than other missiles. Their maneuverability allows them to evade an adversary's missile defenses.[25] If any state produces these missiles in large quantities, it could destroy another state's nuclear arsenals before the victim has a chance to respond.

Hypersonic missiles do not nullify the doctrine of MAD since, in theory, the United States has approximately half of its nuclear arsenal on *Ohio*-class submarines, which are difficult to track or destroy. However, it is fair to say hypersonic missiles provide a significant advantage for any country that can deploy them in high numbers.

Currently, Russia claims it is the first to deploy a "regiment of Avangard hypersonic missiles." According to Defense Minister Sergei Shoigu, the "Avangard hypersonic glide vehicle entered service at 10:00 Moscow time on 27 December [2019]."[26] President Vladimir Putin claims the nuclear-capable missiles can travel more than twenty times the speed of sound, which if true puts Russia ahead of other nations.

Russia's actions are likely in response to the U.S. Navy's successful test of a prototype hypersonic missile in November 2017, but experts suggest the navy's deployment of a hypersonic missile is still years away.[27] Also, there is a significant difference between the goals of the United States and Russia. According to the *Verge*,

> James Acton, co-director of the Nuclear Policy Program and a senior fellow at the Carnegie Endowment for International Peace, was more worried, calling Putin's remarks "really, really bad" on Twitter. "Over the last couple of decades, the U.S. and Russia have not been focused on the same thing," he said. "The U.S. has been focused on non-nuclear hypersonic munitions, while Russia has been focused on nuclear-armed hypersonic munitions."
>
> Since a nuclear weapon is more destructive than a non-nuclear weapon, Acton said, the mechanism for delivery doesn't need to be as precise; it can miss by hundreds of yards or more and still obliterate its target. A non-nuclear weapon, on the other hand, needs to hit its target within a matter of yards.[28]

While the United States has been silent on adding a nuclear payload to its hypersonic missile, experts believe that it would be relatively easy to accomplish once it is operational.[29]

6. Automation: Advanced nations, such as the United States, Russia, and China, are working to harness automation—specifically, AI and robotics—for military purposes.[30] For example, on January 30, 2018, the U.S. Defense Advanced Research Projects Agency (DARPA) released the first prototype of an autonomous sub-hunting surface ship, christened *Sea Hunter*, to the Office of Naval Research (ONR).[31] According to a January 30, 2018, DARPA press release, "ONR will continue developing the revolutionary prototype vehicle—the first of what could ultimately become an entirely new class of ocean-going vessel able to traverse thousands of kilometers over open seas for months at a time, without a single crew member aboard—as the Medium Displacement Unmanned Surface Vehicle."[32]

While this development aligns with the U.S. Third Offset Strategy, which emphasizes AI and robotics as crucial elements, eventually military automation along these lines could allow states to use large swarms of robotic vessels to find and destroy enemy nuclear submarines. This tactic would potentially destabilize the doctrine of

MAD and dramatically raise the risk of nuclear war. For example, if Russia thinks the United States is capable of neutralizing its nuclear deterrent, then it may take drastic action to preempt that outcome. Such action could mean using its nuclear weapons early in a conflict to frighten the United States into backing down. Although this may all sound theoretical, mounting evidence shows that this is what some Russian strategists and possibly Chinese ones are planning.[33]

Although no state currently has the power to wage nuclear war with a capable adversary without fear of annihilation, the six technological evolutions discussed in this chapter could disrupt the doctrine of MAD. Any disruption to MAD could lead to tilting the balance of terror. Unfortunately, the United States, Russia, and China, as well as a handful of other nations, are quickly working to gain a lead in one or more of these six technologies. As a hedge against Armageddon, the United States is investing heavily in all six technologies, work that would disrupt the doctrine of MAD and enable the United States to use nuclear weapons without fear of reprisal.

How close is any nation to gaining dominance in one or more of these six technologies and disrupting the doctrine of MAD? This question takes us to chapter 2.

2

The Quest for Global Dominance Using Conventional Weapons

Now, for the moment, we are safe. The only kind of international
violence that worries most people in the developed countries is
terrorism: from imminent heart attack to a bad case of hangnail in
fifteen years flat. We are very lucky people—but we need to use the
time we have been granted wisely, because total war is only sleeping.
All the major states are still organized for war, and all that is needed
for the world to slide back into a nuclear confrontation is a twist of
the kaleidoscope that shifts international relations into a
new pattern of rival alliances.

—GWYNNE DYER

We are living in a period that historians term the Long Peace. By
this, they mean that since the end of World War II, there has
not been a full-scale war between great powers. We should
count that as one of our blessings; however, it has also induced a
kind of amnesia. Following World War II, most people throughout
the world had vivid memories of the horrific destruction wreaked by
nuclear weapons. During the Cold War between the United States
and the Soviet Union, those memories remained foremost in people's
psyche. Then a strange, unexpected event unfolded: the USSR collapsed. From that point, the United States defined the world order,
and people throughout the world began to forget.

To some extent, they may have willfully erased the dreadful consequences of a full-scale thermonuclear war. Russia, however, still

27

has a formidable nuclear weapons capability. China, whose economy is second only to that of the United States, is rapidly investing in its military and has its own impressive nuclear weapons capability. Unfortunately, as of January 23, 2020, taking all these developments into consideration, the *Bulletin of the Atomic Scientists* has decided to set its "doomsday clock" to "100 seconds to midnight," the closest to catastrophe it has been since 1953.[1]

This symbolic setting of the doomsday clock suggests that the use of nuclear weapons is more likely now than at any time since the Cold War. To my mind, the most likely reason for the erosion of the global security situation is due to mass amnesia regarding the horrific devastation of nuclear weapons. The *National Interest*, an American international affairs magazine, observed, "The worst mistake to make about nuclear weapons is to believe that they are ordinary arms, available for military use like any other."[2]

The torch of national leadership is now in the hands of a generation whose only understanding of the horrendous use of nuclear weapons during World War II is from history books. Given the tensions between the United States and China regarding the latter's disputed claims over the China Sea and Taiwan, as well as issues with North Korea, Iran, and Russia, the possibility of a conflict is increasing. Unfortunately, such a conflict could go nuclear, intentionally or unintentionally, and the world appears complacent about the horrors of a potential nuclear war. Some in the media even suggest we could win a nuclear war. For example, here is a 2018 headline from the *Hill*, an American political newspaper and website: "Only Trump Can Restore America's Ability to Win a Nuclear War."[3] This apparent widespread ignorance regarding nuclear conflict is forcing me to ask, Is nuclear war winnable?

Is a Nuclear War Winnable?

It has been well over seventy-five years since the United States dropped nuclear weapons on the Japanese cities of Hiroshima and Nagasaki during World War II. Although they were small nuclear weapons by today's standards, their combined death toll reached over 250,000 people, most of whom were civilians.[4]

The bombs dropped on Hiroshima and Nagasaki were atomic bombs, meaning they derived their destructive force from nuclear fission. As noted in chapter 1, nuclear fission results when an atom splits into two or more fragments, releasing its binding energy. A nuclear explosion occurs when a significant amount of fissionable material is present, and the fission of one generation of nuclei produces particles that cause the fission of at least an equal number of nuclei of successive generations. In the parlance of nuclear physics, this is a chain reaction, where atoms split, releasing subatomic particles that split more atoms until the process consumes the fissionable material. Because a nuclear weapon typically has pounds of fissionable material, an enormous number of atoms split and enable the chain reaction to release a massive amount of energy, including ionizing radiation, such as neutrons, gamma rays, alpha particles, and electrons moving at speeds close to the speed of light. This released radiation explains why many people who were not directly killed by the atomic bomb's blast later died from radiation poisoning.

The U.S. military describes the destructive force of nuclear weapons in terms of tons of TNT. For example, the U.S. military describes the Little Boy atomic bomb dropped on Hiroshima on August 6, 1945, as equivalent to about fifteen kilotons (fifteen thousand tons) of TNT, and the Fat Man atomic bomb dropped on Nagasaki on August 9, 1945, as equivalent to about twenty kilotons of TNT. However, this convention of expressing the energy released in terms of TNT does not change one important fact: a nuclear explosion accomplishes its devastation via the release of energy, which is typically measured in joules. One ton of TNT is a unit of energy equal to 4.184 gigajoules.[5] (A gigajoule is equal to a billion joules. An easy way to think about it is that a watt is the flow of one joule per second of energy. Your ordinary household 100-watt light bulb emits 100 joules per second.) For completeness, we can express the release of the Little Boy as follows: 15 tons × 4.184 gigajoules

= 62.76 gigajoules, or 17,433,333.3 watt-hours. Similarly, Fat Man equals 83.68 gigajoules, or 23,244,444.4 watt-hours. I believe the U.S. military uses the "ton of TNT" convention because it is easier to grasp. However, as the military moves to directed-energy weapons, which are able to replicate the destructive power of nuclear weapons, I expect it will describe their destructive capabilities in terms of energy. If I am right, the word "joules" will become common in the military vernacular.

At present, not only has the number of nuclear-armed nations expanded—to nine—but nuclear weapons technology has also evolved. Instead of fission nuclear weapons, we now have thermonuclear weapons, often called hydrogen bombs. In a hydrogen bomb, fission is only the beginning of the process. The initial energy from the fission process ignites a fusion reaction between the hydrogen isotopes deuterium and tritium, which fuse to form helium and release energy. With sufficient quantities of deuterium and tritium, a chain reaction results in releasing an enormous amount of energy that is thousands of times more potent than the first atom bombs used in World War II. Instead of talking in kilotons of TNT, the world militaries describe these bombs in megatons (i.e., millions of tons) of TNT. For example, the B83—the U.S. military's current gravity bomb (a free-falling bomb dropped from a plane)—is a thermonuclear weapon with an adjustable yield, or destructive power, ranging from the low kilotons up to a maximum of 1.2 megatons.[6] To put this in perspective, in Hiroshima and Nagasaki, each bomb dropped was less than twenty-one kilotons of TNT and destroyed almost everything within a one-mile radius from the center of the explosion.[7] The total death toll, as previously noted, exceeded 250,000 thousand people, which amounted to about half of the population of Hiroshima and Nagasaki.[8] If the United States had dropped 1-megaton bombs on Hiroshima and Nagasaki, the blast radius of each bomb would have been over three miles, and the death toll would have been over half a million people, encompassing the entire population of each city in 1945.[9]

As of this writing, there are nearly 14,000 nuclear weapons in the world. The United States has 6,550, and Russia, 6,800. The remainder by country in ascending order is found in North Korea (10–20), Israel (80), India (120–130), Pakistan (130–140), United Kingdom (215), China (270), and France (300).[10] These numbers are approximations since each country holds its exact number as a tightly held secret. With these figures in mind, let's address two questions that explore the limits of nuclear war.

What Would Happen If the United States and Russia Engaged
in a Full-Scale Nuclear War?

According to an Office of Technology study for the U.S. Congress in 1979, in the event of a nuclear war, 70 million to 160 million people (35–80 percent of the population) would immediately die in the United States.[11] Russian fatalities would be approximately 20–40 percent lower. Many more in both countries would die from injuries, cancer-related deaths, and psychological trauma. In updating these numbers to reflect the populations as of 2018 and the urbanization of the United States, the death toll is certainly higher. In 1979 the United States had a population of about 225 million; in 2018 it was 327 million, with much of the growth occurring in urban areas, where approximately 80 percent of the U.S. population now lives.[12] By contrast, Russia's population has only grown modestly, from a population of about 137 million in 1979 to 144 million in 2018.[13]

On the surface, even with the increased urbanization of the United States, the report's numbers appear to suggest that even the combatant nations could survive a full-scale nuclear exchange. However, that is not the case. In addition to the immediate deaths and the total destruction of cities by nuclear blasts, more casualties would occur from the potential aftermath of a nuclear war: firestorms, widespread radiation sickness from the bombs and radioactive fallout, the loss of modern technology due to electromagnetic pulses, and a nuclear winter resulting in worldwide famine.

Deaths from the nuclear blast, firestorms, and radiation are relatively easy to grasp. Deaths from the effects of an EMP blackout and a nuclear winter are more challenging to understand. Therefore, let us discuss each.

EMP Blackout: A nuclear detonation causes an electromagnetic pulse, which produces rapidly varying electric and magnetic fields. In turn, those fields cause electrical and electronic systems to experience damaging current and voltage surges, resulting in a blackout. (*Note*: In physics, a current generates a magnetic field, and a rapid change in a magnetic field generates a surge current.) How severe would an EMP blackout be? EMP expert Peter Vincent Pry concludes in a 2017 report that in a widespread EMP attack, "nine of 10 Americans" would die "from starvation, disease, and societal collapse."[14]

Reading that last line is chilling. Even if portions of the country are not affected by the nuclear blast, radiation, and firestorms, the United States of America would cease to exist if the effects of EMPs caused the deaths of 90 percent of its residents. According to Pry, Russia views EMP differently—as a "revolution in military affairs."[15]

Nuclear Winter: If you are fortunate enough to survive the nuclear blast, radiation, fallout, and EMP blackout, you are still likely to perish in the coming years. Alan Robock and Owen Brian Toon, in their paper, "Self-assured Destruction: The Climate Impacts of Nuclear War," hypothesize that a thermonuclear war could result in a nuclear winter that would be the end of all modern civilization on Earth.[16] The nuclear winter would result from the smoke and soot arising from burning wood, plastics, and petroleum fuels in nuclear-devastated cities and blocking the sun's rays. A recent study reports the resulting darkness would cool the core farming regions of the United States, Europe, Russia, and China by about 54–68 degrees Fahrenheit.[17] The cooling effect would reduce crop yields and, due to disrupted agricultural production and distribution, lead to a "nuclear famine" characterized by mass starvation.[18] The simple takeaway is modern civilization on Earth would cease to exist, and the remnants of humanity would find themselves struggling to survive.

Conclusion: There would be no winners in a full-scale nuclear exchange between the United States and Russia.

At first, you might assume that a limited, regional nuclear war is winnable. On the surface, it would appear to be a reasonable assumption given over five hundred atmospheric nuclear explosions (including eight underwater) with a total yield of 545 megatons have been conducted since 1945 and did not lead to a nuclear winter. You may wonder why.

As previously discussed, a nuclear winter is the result of smoke (i.e., black carbon) in the stratosphere blocking the sun. With nuclear testing, we did not experience a nuclear winter because nations conducted the aboveground tests in areas where there was not much to burn. For example, the United States carried out nuclear testing in the Nevada desert and did not generate any significant amount of smoke because the area lacked the combustible materials found in urban areas.

In a limited, regional nuclear war, adversaries likely would target each other's military assets and command-and-control operations, both of which could be near or within cities. Indeed, adversaries may target large population areas as a tactic to demoralize the populace. Now let us address the question of whether this limited, nuclear exchange would lead to a nuclear winter.

A 2014 paper by Michael J. Mills and colleagues studied the global impacts of a regional nuclear war.[19] Their model assumed a limited, regional nuclear war between India and Pakistan in which each side detonated fifty fifteen-kiloton nuclear weapons, producing about five teragrams (5 million metric tons) of black carbon. In addition, they assumed the black carbon would self-loft to the stratosphere, where it would spread globally. What impact would this have on the earth? Their calculations "show that global ozone losses of 20%–50% over populated areas, levels unprecedented in human history, would accompany the coldest average surface temperatures in the last 1000 years. . . . Killing frosts would reduce growing seasons by 10–40 days per year for 5 years. Surface temperatures would be reduced for more than 25 years due to thermal inertia and albedo effects in the ocean and expanded sea ice. The combined cooling and enhanced uv [ultraviolet] would put significant pressures on global food supplies and could trigger a global nuclear famine."[20]

Conclusion: Even a limited, regional nuclear war could result in a nuclear winter and worldwide famine, which would threaten humanity's survival.

After examining these two questions, which represent the extremes of nuclear war, we can conclude that no nuclear war is winnable. There has been significant research in this area, and we have covered the key findings. It is reasonable to believe the nations with nuclear weapons understand that using them in any capacity risks global human annihilation. As a result, nuclear powers such as the United States, Russia, China, and others are working to increase the potency of their conventional weapons to further their international agendas. As they do so, however, they are also increasing the pace of warfare.

Increasing the Pace of Warfare

In chapter 1, we discussed six military technologies that have the potential to tilt the balance of terror. One common denominator running through them is their ability to increase the pace of warfare, from hypersonic missiles traveling at five times the speed of sound to laser weapons that travel at the speed of light (three thousand times the speed of sound). Given this insight, it is reasonable to ask, Why are the major military powers, such as the United States, China, and Russia, focused on increasing the pace of warfare? To address this question, let us look at it from two perspectives—strategy and technology.

Strategically, speed in warfare plays a critical role. For example, the U.S. military must be able to respond to rapidly changing situations. Therefore, its decision-making needs to exceed any adversary's military initiative in three ways:

1. By anticipating an adversary's potential action and having a plan to counter it

2. By having better mobility than an adversary

3. By providing real-time voice, data, and imagery information via satellite to every level of command, including leaders at the lowest levels in the military hierarchy

These capabilities radically shorten what military strategists call the decision space. If we look at World Wars I and II, the United States had time to marshal its immense strength. For example, it took nearly a year after the bombing of Pearl Harbor on December 7, 1941, for those American troops who arrived in the British Isles in January 1942 to see action against the Axis powers.[21] Today's adversaries and the pace of war no longer afford that time to react to acts of war that happen suddenly and without provocation. According to Gen. Joseph F. Dunford Jr., former chairman of the Joint Chiefs of Staff, "The character of war in the 21st century has changed, and if we fail to keep pace with the speed of war, we will lose the ability to compete."[22] Although this example views speed through the eyes of the U.S. military, I believe it applies to our most capable adversaries as well.

Next, let us examine how technology is changing the pace of warfare. World militaries are seeking technological advantages that enable them to deliver debilitating force at speeds that do not allow an adversary to take countermeasures. While on the surface this may seem like a new tactic, it is quite old. For example, Chinese military strategist and philosopher Sun Tzu's *The Art of War*, which dates to about 500 BCE, has this to say about speed in warfare: "Speed is the essence of war. Take advantage of the enemy's unpreparedness; travel by unexpected routes and strike him where he has taken no precautions."[23]

This strategy played a role in numerous historical battles. For example, we see its implementation during the American Revolutionary War. Schoolchildren typically learn that Gen. George Washington and his revolutionary army crossed the icy Delaware River on the night of December 25, 1776, to launch a surprise attack against the Hessian forces in Trenton, New Jersey. On December 26, 1776, General Washington defeated a garrison of Hessian mercenaries in the Battle of Trenton. Many consider this battle a turning point in the Revolutionary War.

You may argue General Washington used surprise, not speed, to achieve victory. That is a fair point. However, now in the age of satellite surveillance, speed is what enables the element of surprise in war. Any prelude to an attack, such as amassing forces within striking distance of an adversary, is easily detectable via satellite data. Current military thinking is to use speed so that an adversary has no time to prepare for an attack. This thinking thus underpins hypersonic missile development. The United States, China, and Russia have tested hypersonic missiles, and it appears that both China and Russia lead the United States in this technology. Reuters, an international news organization, reported, "Admiral Harry Harris, the former head of U.S. Pacific Command, told the House Armed Services Committee in February last year [2018] that hypersonic weapons were one of a range of advanced technologies where China was beginning to outpace the U.S. military, challenging its dominance in the Asia-Pacific region."[24] CNBC, an American business news cable TV channel, also reported, "Putin, who was speaking at a forum in the Black Sea resort of Sochi [in 2018], added that Moscow's hypersonic weapons program was ahead of its competitors."[25]

As a result, the United States is making hypersonic missile development a top priority. Phys.org, a science, research, and technology news aggregator, observed: "The Pentagon has declared hypersonics to be its number one research and development technical priority. The president's recent budget request proposes allocating almost $3 billion to develop hypersonic weapons and defense systems against potential adversaries' hypersonic weapons."[26]

In addition, the United States is developing laser weapons, which are in their initial deployment by the U.S. Navy.[27] Based on Einstein's theories of relativity, nothing in the universe can travel faster than light; therefore, even a hypersonic missile would appear to be standing still compared to the speed of light projected by an antimissile laser weapon. Although the United States does not openly state a plan to use lasers against hypersonic missiles, I believe this would be among its ultimate goals. For now, though, imagine a scenario where the U.S. military has a mirror orbiting in space. Then imagine that it could use a powerful laser to bounce a beam off that mirror to destroy any target on Earth, in space, or during flight, such as a hypersonic missile. You may think this idea is the stuff of a science fiction movie, but the laser development at Lawrence Livermore National Laboratory may make it a reality (see chapter 4). Conceivably such a weapon could destroy an adversary's entire military capability in a flash, pun intended. Even a nation's ability to launch a counterstrike via missiles from submarines might become futile if a laser can destroy them in flight. Conventional antimissile weapons require calculating the trajectory of an incoming missile; the speed of a laser makes this unnecessary. A laser fired at a missile's radar coordinate would hit it as if suspended in flight. That is one of the critical advantages lasers offer as antimissile defense weapons.

Directed-energy weapons hold the promise of conventional military domination. By most accounts, the United States leads the world in this military technology. This knowledge may be reassuring. On the surface, it appears that with this technology, the United States will achieve conventional weapons dominance. However, history teaches us military secrets are hard to keep. For example, the United States once had the lead in nuclear weapons, using its first weapons on Japan in 1945 to effectively end the Pacific war, but by 1949 the Soviet Union had acquired nuclear weapons technology via espionage. This raises a significant question: Why are military secrets so hard to keep? Let us digress for a moment to address it.

The Difficulty in Keeping Military Secrets

The fundamental reason military secrets are hard to keep is that the development of new weapons requires the work of many people, typically thousands. As Benjamin Franklin wrote in *Poor Richard's Almanack* in 1735, "Three may keep a secret if two are dead." While this may seem extreme, there is more than a kernel of truth in his axiom. To illustrate, here is a simple example from the Federation of American Scientists: "The Manhattan Project to develop the first atomic bomb during World War II was among the most highly classified and tightly secured programs ever undertaken by the U.S. government. Nevertheless, it generated more than 1,500 leak investigations involving unauthorized disclosures of classified Project information."[28]

In light of all these leaks, Gen. L. R. Groves Jr., head of the Manhattan Project, told Congress, "I would like to say that the only thing that would preserve security would be to lock everybody up, and when they decided to leave to shoot them and be done with it. That is the only way you could have perfect security."[29]

General Groves was making a point: it is impossible to have perfect security when thousands of people are working on a classified program. Unfortunately, building advanced military weapons requires many people, including those who work in private industry. For example, I held a secret clearance and worked hand in glove with members of the U.S. military to develop new weapon capabilities, as did hundreds of others who worked under my management. This type of working relationship is not unique. It is typical.

As sophisticated U.S. adversaries steal classified information via cyber warfare, I believe it is impossible to maintain security over any protracted length of time. Thus, what is essential in achieving military superiority is maintaining relative momentum—that is, staying years ahead in deploying weapons against which an adversary has no defense. To my mind, this is where things stand today regarding directed-energy weapons, with the Department of Defense collaborating with industry to lead in developing and deploying them. This union of military and industry is, in the common vernacular, an example of the military-industrial complex.

Inside the Military-Industrial Complex

The first time I heard the phrase "military-industrial complex" was during President Dwight Eisenhower's farewell address to the nation on January 17, 1961. Here is an excerpt: "In the councils of government, we must guard against the acquisition of unwarranted influence, whether sought or unsought, by the military-industrial complex. The potential for the disastrous rise of misplaced power exists and will persist. We must never let the weight of this combination endanger our liberties or democratic processes."[30] In essence, President Eisenhower seized the opportunity of his final address to warn the nation of a threat to democracy—namely, an overly powerful, politically motivated military-industrial complex.

The threat may seem elusive and nonspecific, and it raises the questions of what the military-industrial complex is and who is involved. Most may answer it is a network of companies that make military weapons. An obvious example would be Lockheed Martin—along with subcontractors Northrop Grumman, Pratt & Whitney, and BAE Systems—as it manufactures the F-35, the fifth-generation stealth multirole fighter aircraft. Media coverage of the F-35 continues to be enormous, and many in the world community know about it. However, providing a complete list of the companies engaged in making any military weapon is extremely difficult. Many critical military technologies remain unknown to both the media and the public. Let me give you an example from my experience.

In 1980 I was the manager of sensor development and manufacturing for Honeywell. Among the sensors we produced were integrated circuit magnetic sensors, typically about the size of your pinky fingernail. Briefly, this sensor can discern magnetic fields. Applications ranged from detecting when someone pressed a specific key on a keyboard to measuring anomalies in the earth's magnetic field. In the course of business, we received an order for a uniquely designed magnetic sensor from another Honeywell business unit in Europe. To me, this request was business as usual. We supplied magnetic sensors to numerous Honeywell business units, and nothing about this order appeared unusual. We designed the sensor to the purchase order's specifications and provided the completed products to our shipping department for shipment. However, a few days later, the shipping department asked me to sign an International Traffic in Arms Regulations (ITAR) form requesting permission from the U.S. government to ship armaments of war. I paused and thought, *What armaments of war?*

Nothing about the magnetic sensors resembled a weapon. Being cautious, I sought to understand more about the application of sensors and the ITAR document before signing. After communicating with my Honeywell client in Europe, I discovered that the magnetic sensors were a critical element of an anti-tank mine, a type of land mine designed to destroy enemy tanks. Each vehicle from a truck to a tank distorts the earth's magnetic field in a specific way, and that is its magnetic signature. The magnetic sensors we provided were able not only to sense a tank but also to determine if its magnetic signature was that of an adversary's tank. With this capability, the mine would only destroy an adversary's tank. As the duly authorized Honeywell representative, I signed the ITAR document. As I signed, I realized that from that moment, we were part of the military-industrial complex.

I relate this story to make two points:

1. The military-industrial complex is hard to define and includes companies typically not identified as defense contractors.

2. The military-industrial complex is much larger than the Department of Defense and its budget.

The reasons for these points are subtle. Many products and technologies play dual roles. For example, the microprocessor in your computer can simulate a war game or be a vital component in a weapon of war. Thus, the DOD budget does not reflect the development of such products; it only includes their acquisition costs.

In 2018 the U.S. military budget amounted to roughly 3.2 percent of the economy.[31] About a third (1 percent) is for weapons development and deployment, and much of what the U.S. military needs to develop weapons typically rests on the collaboration of academia and industry. Having worked in the defense industry and having bid on defense contracts, I can attest that bidding on a typical defense contract requires having a credible capability in place that is applicable to the contract. In simple terms, at Honeywell we built on the technology we already had and were not starting from scratch. Otherwise, our ability to execute the contract would have been questionable, and our bid would not have been competitive. Thus, the one-third of the U.S. military budget for weapons acquisition does not reflect the investments companies make to position themselves to bid on DOD contracts.

The bottom line is that the military-industrial complex is much larger than the acquisitions reflected in the 1 percent of the U.S. defense budget associated with weapons procurement. Industry and academic institutions privately fund much of the fundamental technology regarding weapons development.

Why is this important? It demonstrates that the economy of a country matters for two strategic reasons:

1. Only a country with a strong economy will have the industrial and academic institutions with the necessary discretionary funding to finance the essential technology that will position them to participate in weapons procurement.

2. Only a country with a strong economy can sustain a large defense budget as it procures the latest weapons. Recall that the Soviet Union collapsed in late 1991 after participating in a costly arms race with the United States.

Our most capable potential adversary, China, can have a defense budget that matches that of the U.S. military. While on the surface, the U.S. government significantly appears to outspend China on defense, a complete analysis reveals that is not the case. In 2017, for example, China spent about the same on defense as the United States. I recognize that, in the words of Carl Sagan, "extraordinary claims require extraordinary evidence." Thus, appendix A of this volume adjusts the defense budget of China and the United States for purchasing power parity—a measure of how much a particular currency will buy in the respective regional market relative to another currency—and labor costs. The results in appendix A yield a stark new reality: the defense-spending strategy of potential adversaries plays a more significant role than it has historically and especially in the new arms race, which includes the race for directed-energy weapons.

A new revolution is coming in warfare, and, if you will pardon the pun, it is coming at the speed of light. This revolution concerns the emergence of directed-energy weapons, which promise to be even more potent than nuclear weapons. If you are skeptical about this statement, let me relate a scenario involving a laser as a weapon to provide insight into the potency of directed-energy weapons.

Imagine a land-based laser capable of generating the power equivalence of a nuclear weapon. Now imagine bouncing the laser's beam off a satellite's mirror and reflecting the beam to hit a target anywhere on Earth. What would its effect be on the target? The target would become plasma, the fourth state of matter, a jumble of excited nuclei, atoms stripped of their electrons. All life and structures within the diameter of the laser beam would be destroyed. Next, imagine the target is the city of Beijing. In an instant, 21.5 million people would die. In another fraction of a second, the laser could attack another target; in a minute, it could destroy China. I only used Beijing and China as an example. The same scenario would hold for New York City and the United States, or for any city and any country.

If you are incredulous, let me give you some facts. Numerous countries around the world are making powerful lasers for military purposes, and their ultimate goal is to make them as powerful as a nuclear weapon. Although we have a long way to go, the United States is actively building the world's largest continuous laser at the National Ignition Facility (NIF) of the Lawrence Livermore National Laboratory (LLNL).[32] The LLNL website states, "NIF enables scientists to create extreme states of matter [using a laser], including temperatures of 100 million degrees and pressures that exceed 100 billion times Earth's atmosphere. NIF supports national security, fundamental science, energy security, and national competitiveness missions."[33]

In simple terms, the NIF laser duplicates the phenomena that occur in a nuclear weapon, but to date, it is only able to generate a tiny fraction of the energy output of the smallest nuclear weapon.[34]

The facility's long-term goal is to create a fusion reaction that produces a greater energy output than the laser's input. The implication is this capability would serve as a power generation plant, essentially making all other types of power generation obsolete. Although the Lawrence Livermore National Laboratory's National Ignition Facility does not discuss the laser's weapon applications, its national security mission suggests it is actively pursuing weapon applications for this laser. Based on this information, plans for the preceding scenario I provided may be in development. As mentioned earlier, I also believe the LLNL laser may serve as an antimissile weapon, especially since the United States is scrambling to develop defense systems against its potential adversaries' hypersonic weapons.[35]

One aspect of a laser weapon is that, similar to the *Star Trek* phasers, it can be "dialed in"; in other words, its power is adjustable from low to high. On low, it might dazzle an enemy combatant, causing temporary blindness or minimal damage to an enemy drone, for example. The damage may be sufficient to cause the drone to malfunction, allowing its recovery for examination. On high, it might kill the enemy combatant; similarly, it might destroy a drone, fighter jet, or missile. The laser weapon deployed by the U.S. military has this dial-in capability.[36]

Why Nations Are Pursuing Directed-Energy Weapons

Since the U.S. and other militaries already have weapons that can duplicate the destruction of lasers and other directed-energy weapons, you may wonder why world militaries are pursuing directed-energy weapons. Here are the fundamental reasons that drive this pursuit:

They are conventional weapons, regardless of their devastation capability, and thus are not covered by treaties that apply to nuclear weapons.

Unlike a nuclear weapon, no radiation is released during their use.

Unlike most military weapons, directed-energy weapons are silent, and the beams they project are invisible, providing a stealth attribute.

Directed-energy weapons, such as lasers, have a nearly flat trajectory and are unaffected by windage. However, they can be affected by atmospheric conditions (see chapter 4).

In space, directed-energy weapons would be unaffected by Earth's atmosphere, which makes them ideal for space warfare.

They are cost-effective. A laser beam, for example, can take out a drone for about a dollar a shot.

The technology for directed-energy weapons is rapidly advancing, as is evident when examining the laser deployed by the U.S. Navy. It is relatively small and looks as though it came out of a *Star Trek* movie (see figure 1).

Ultimately, directed-energy weapons can be a game changer. Conceivably, by 2030, they will be capable of destroying ballistic missiles, hypersonic cruise missiles, hypersonic glide vehicles, and swarm drone and missile attacks.

Concluding Thoughts

Any significant conflict involving nuclear weapons would threaten the survival of humanity. As a result, nuclear powers such as the United States, Russia, and China are working to increase the potency of their conventional weapons to further their international agendas. They are essentially playing a game of cat and mouse on an international level, leading to a new arms race with directed-energy weapons becoming its key elements. Based on publicly available information, the United States appears to be leading in the development and deployment of directed-energy weapons. While these weapons offer enormous potential to dominate the military landscape, their true capabilities and latest developments are closely guarded secrets.

As a result, I believe that directed-energy weapons are in development but hidden from the public. I make this statement knowing that much of my work, going back several decades, still remains secret— for example, my work on our most sophisticated communication satellites in the Milstar (Military Strategic and Tactical Relay) sys-

1. U.S. Navy laser weapon system. Photo by John F. Williams. Courtesy of U.S. Navy.

tem and on one of our most deadly torpedoes, Mark 50, which is the U.S. Navy's advanced lightweight torpedo for use against fast, deep-diving submarines. Although the United States publishes its defense budget, it guards many of its latest weapon developments by designating them as "black programs," meaning knowledge about them requires a top-secret clearance and a "need to know." However, those directed-energy weapons in the public domain serve as a means of plotting their development trajectory. Based on the information in this book, I think you will also conclude that directed-energy weapons will dominate warfare by 2050. The weapons of science fiction franchises such as *Star Trek* and *Star Wars* will be the weapons of world militaries. To my mind, one question remains: What technological strategy will the United States pursue to dominate the twenty-first-century battlefield?

3

The Coming Fourth U.S. Offset Strategy

The supreme art of war is to subdue the enemy without fighting.

—SUN TZU

I recognize understanding the U.S. military is extremely difficult. Much of the difficulty is intentional. The U.S. military is the best-trained, best-prepared, and most technologically sophisticated military in the world, and a veil of secrecy surrounds its strategies and weapons. Indeed, the U.S. military may make public statements that are deliberately misleading to confuse and confound potential adversaries. Unfortunately, those statements also confuse and confound Americans.

Is there any way to understand the U.S. military's strategies and weapons technology? The answer is yes, and it has to do with its offset strategies. The U.S. military uses this term in a specific official context—that is, to refer to weapons and technologies it pursues to gain an advantage over potential adversaries. As with most militaries, the U.S. military learned the importance of having an offset strategy through the blood, sweat, and tears of war, starting with the Korean War.

Before the war, the U.S. military's strategies and weapons technologies were improvisations to address contemporary and perceived needs. Similar to a three-ring circus, with lions in one ring, elephants in a second, and clowns in a third, the U.S. military lacked a cohesive forward-looking strategy. If you think this view of the U.S. military is harsh, consider the U.S. Pacific Fleet's defeat via the Japanese sneak attack on Pearl Harbor on December 7, 1941. At the time, the U.S. foreign policy of isolationism was not focused on defending the country. The National World War II Museum, focused on military history, characterizes the United States after the attack on Pearl Harbor as "ill-equipped and wounded."[1] It had no military-industrial complex. The U.S. Army was smaller than Portugal's, ranking seventeenth in the world.[2] Fortunately, in 1941, the mainland of the United States remained out of easy reach of its new enemies—Japan, Germany, and Italy—however, we were at war and faced significant challenges. According to the National World War II Museum's website, "Meeting these challenges would require massive government spending, conversion of existing industries to wartime production, construction of huge new factories, changes in consumption, and restrictions on many aspects of American life. Government, industry, and labor would need to cooperate. Contributions from all Americans, young and old, men and women, would be necessary to build up what President Roosevelt called the 'Arsenal of Democracy.'"[3]

In a sense, the attack on Pearl Harbor served as a tragic wake-up call. The famous quote attributed to Adm. Isoroku Yamamoto in the 1970 film *Tora! Tora! Tora!* rings true: "I fear all we have done is to awaken a sleeping giant and fill him with a terrible resolve." To build the arsenal of democracy to fight World War II required establishing what President Eisenhower later called the military-industrial complex. The United States emerged from that war as a superpower, the sole country with nuclear weapons, but that distinction was about to change. On August 29, 1949, the Soviet Union detonated its first atom bomb. Almost in the blink of an eye, the United States' former World War II ally became a potential adversary. The United States could no longer embrace isolationism. The short-lived peace crumbled into a Cold War.

The First U.S. Offset Strategy

A by-product of the Cold War was the splitting of Korea into two sovereign states. The government of the Democratic People's Republic of Korea (North Korea) was under the communist direction of Kim Il Sung, while the government of the Republic of Korea (South Korea) was a capitalist state under the leadership of Syngman Rhee.[4] Both governments claimed to be the sole legitimate government of all of Korea. This dispute set the stage for what became the first of many of the Cold War's proxy wars. On June 25, 1950, with the support of China and the Soviet Union, North Korea invaded South Korea, igniting the Korean War.[5] As a result, the newly formed United Nations sent UN forces, led by the United States, to South Korea to repel the invasion.[6]

I was a young child at the time, but I can remember my older cousin going off to fight in the Korean War. It was a tense time in our family, especially after my cousin went missing in action. Fortunately, after several days U.S. forces found him, though he was unconscious and suffering from frostbite. Eventually, he recovered and returned home, no longer able to serve. Although I was young, this incident from the early 1950s remains vivid in my memory. It likely affected other families throughout the United States in a similar way. The nation was still recovering from World War II. No one wanted another war.

On his return from Korea in early December 1952, President-elect Dwight D. Eisenhower, aboard the heavy cruiser USS *Helena* somewhere in the middle of the Pacific Ocean, conferred with several top incoming administration officials. In what speechwriter Emmet J. Hughes described as an "antiseptically spare and cold" conference room, the president-elect and his advisers strategized.[7] Their focus was to end the Korean conflict they had inherited from President Harry Truman. Unfortunately, after witnessing the stalemate firsthand, they knew the geopolitical conflict favored the Soviet Union's overwhelming conventional forces. The United States and its allies were fighting a large-scale war halfway around the world in the USSR's backyard. Eisenhower worried about the drain on U.S. and European resources, as well as the impact on the morale of war-weary Americans and Europeans. Distraught, he sought a strategic solution. During this bleak meeting, Secretary of State–designate

John Foster Dulles advocated using the threat of America's nuclear weapons to deter Soviet-backed expansionism. That resonated with Eisenhower.

Following the group's return to the States and after having three handpicked teams surface strategic options, the Eisenhower administration adopted a new national defense policy, New Look, which emphasized using the United States' superiority in nuclear weapons to offset the Soviets' advantage in conventional weapons. Although the administration called this deterrence strategy New Look, history records it as the First U.S. Offset Strategy.[8] Eventually, the North Korean military and the South Korean military signed an armistice on July 27, 1953—a cease-fire between the military forces but not a peace agreement between the two governments. I believe President Eisenhower's New Look strategy was a crucial factor in securing the truce.

At this point, we must recognize the U.S. government controlled the program of nuclear weapons' development and manufacture. Potential adversaries could not gain access to this technology except via espionage. Although the Soviets succeeded in doing that, in 1953 they still lagged the United States, which had a clear strategic advantage in both the technology and the number of nuclear weapons. Also, it is essential to recognize that an offset strategy needs to both deter war with potential adversaries and win it if necessary. The U.S. offset strategy, in this context, emphasized those military technologies that countered the advantages of its potential adversaries. In the case of the Korean War, New Look—that is, the First U.S. Offset Strategy—brought both sides to the negotiation table and led to an armistice.

The Second U.S. Offset Strategy

One aspect of technology is that it continually evolves, and its adoption spreads. Consider the telephone, for example. Alexander Graham Bell received U.S. Patent 161,739 for the phone in 1876 as an "apparatus for transmitting vocal or other sounds telegraphically." By 1900, there were about 600,000 phones in the United States; by

1910, there were 5.8 million, growing by almost a factor of ten in a decade. You can see from the wording of the patent, especially the word "telegraphically," that the U.S. Patent Office considered the telephone an evolution of the telegraph. However, the development of this technology did not stop there. Initial phones, and many landlines still in use today, required wires to connect the user to the transmitter/receiver device, that device to a switchboard, and the switchboard to another transmitter/receiver device. Fast forwarding to the present, most people have smartphones, wireless devices that do a great deal more than send and receive telephone calls. A smartphone is a handheld computer with more computing power than NASA engineers had when they put a man on the moon. It also incorporates a camera and the ability to send text messages and photographs. The next point on its evolutionary path integrated the smartphone with a watch. Will the evolution stop here? Unlikely, but that is where we are as of this writing.

This short history of the telephone's development and adoption illustrates how quickly technology evolves and spreads. With this understanding, let us return to offset strategies and their role in advancing military technologies. Similar to all technologies, the offset strategies also evolved and spread from the First U.S. Offset Strategy.

Meanwhile, the nuclear technologies of that first strategy developed. The first early nuclear weapons relied solely on a fission chain reaction. In the next-generation nuclear weapons, the fission chain reaction became the first step in causing a fusion chain reaction, giving rise to more powerful thermonuclear weapons. Soon, engineers discovered how to reduce the size of thermonuclear weapons and deliver them to their targets by intercontinental ballistic missiles. This evolution took time and the work of thousands of people. Even with all this development, however, starting in the 1960s, U.S. dominance in nuclear weapons began to diminish. The technology had spread to other nations.

By the mid-1970s, six nations had nuclear weapons: the United States, Soviet Union, United Kingdom, France, China, and Israel.[9] As the Cold War raged between the United States and the USSR, with near parity in nuclear weapons, it became increasingly apparent that the First U.S. Offset Strategy no longer afforded the United States a technological military advantage over the Soviet Union. Additionally, Warsaw Pact forces outnumbered NATO forces by three to one in Europe. Thus, the United States sought new strategies to regain its technological advantage.

Starting in the 1970s, under President Jimmy Carter, Defense Secretary Harold Brown directed the Department of Defense to begin investing in extended-range, precision-guided munitions, stealth aircraft, and space-based intelligence, surveillance, reconnaissance, communication, and navigation platforms.[10] In 1980 Ronald Reagan's administration continued to fund these technologies, many of which came to the public's attention during the First Gulf War in 1991. While the U.S. government did not officially label its development of these technologies, it became known as the Second U.S. Offset Strategy. In 1991 these technologies clearly gave the United States a distinct military advantage against its potential adversaries. With the dissolution of the Warsaw Treaty Organization on July 1, 1991, and the collapse of the USSR on December 25 of that year, the United States emerged as the only superpower.

Just as technologies evolve, going from prominence to obsolescence, the fortunes of nations also change, rising and falling. As the world's only military superpower, the United States—after the 2001 terrorist attacks (9/11) on New York City, Pennsylvania, and the Pentagon—engaged in fighting global terrorism. While this was an import aspect of ensuring its national security, by concentrating its forces on this fight over the next two decades, the United States lost sight of the changing world.

In 2014 the *Diplomat* reported:

> In a speech at the Reagan National Defense Forum (and accompanying memo), Defense Secretary Chuck Hagel announced the Pentagon's new "Defense Innovation Initiative."
>
> As Hagel explained in the speech, "While we spent over a decade focused on grinding stability operations, countries like Russia and

China have been heavily investing in military modernization programs to blunt our military's technological edge."[11]

Hagel's speech marked the end of the Second U.S. Offset Strategy, which some military historians characterized as the "American Way of War."[12]

The U.S. government collaborated with industry (and built the military-industrial complex) and developed the technologies under the Second U.S. Offset Strategy. Most of these technologies, such as extended-range, precision-guided munitions and stealth aircraft, had no commercial application. As such, they were closely held secrets and not available in the open market. This is not the case in the Third U.S. Offset Strategy.

The Third U.S. Offset Strategy

As discussed in the introduction, the Third U.S. Offset Strategy was front and center from November 15, 2014, to the inauguration of Donald Trump as the forty-fifth president of the United States on January 20, 2017. During that time, it became a buzz phrase to justify "new" weapons and organizations; however, note that the U.S. military does not announce any technology or weapon unless it is already one or two generations ahead of whatever the military reveals. Here the path to the Third U.S. Offset Strategy was something out of a James Bond novel, complete with a secret organization and futuristic weapons.

In 2012 Barack Obama's administration directed the Pentagon's senior leadership to establish a secret new office, the Strategic Capabilities Office (SCO), whose mission was to work on state-of-the-art weapons ranging from swarming micro drones to hypervelocity projectiles.[13] The SCO remained hidden from the public until Secretary of Defense Ash Carter unveiled his 2017 budget plan in a 2016 speech to the Economic Club of Washington.[14] To the surprise of many, he pulled the curtain back on the shadowy SCO, which shared the same building as the Defense Advanced Research Projects Agency.

You may wonder, as I did, why the Obama administration felt the need to create another DARPA-like organization. According to

Defense News, a global website and magazine about politics, business, and technology of defense,

> Although they share some DNA, the SCO's mission is different from that of DARPA. Whereas the latter is focused on finding and prototyping the game-changing technologies for the future fight, the SCO is trying to understand current, existent needs and address them in new ways.
>
> The most public example of an SCO project is the Standard Missile 6, which [the SCO] helped turn from a defensive weapon into a ship-killing one. It's about taking current capabilities that exist and finding new ways to utilize them.[15]

Having worked with DARPA on numerous projects, the difference between it and the SCO became immediately clear to me. Let me explain. As noted previously, while serving as Honeywell's director of engineering focused on integrated circuit sensors, I worked on some highly advanced sensor technologies for the Department of Defense. In some cases, that meant working with DARPA, Honeywell's partner on some of those technologies. At the time, I was the youngest director of engineering in Honeywell, and given my experience in developing commercial sensor products, I knew little about working with the DOD and nothing about working with DARPA. Fortunately, Honeywell had a corporate sales organization to help its executives work with DARPA.

I can vividly recall some elements of my first visit to DARPA in the early 1980s (DARPA moved to its new facility in 2012). At the time, DARPA's headquarters resided at 3701 North Fairfax Drive in Arlington, Virginia. I always had pictured DARPA as having thousands of employees in a temple of doom surrounded by barbed wire fences and with armed guards around the perimeter. To my surprise, its headquarters was just another office building, but it did have a high-security entrance. The corporate sales representative who accompanied me had set up meetings to introduce me to the key people involved in my projects, including the then director of DARPA, Dr. Robert S. Cooper. My meeting with him was short and cordial. Although I was young with little experience working with the DOD, he respected my position in Honeywell and took time from his busy schedule to explain how DARPA operated. I had several takeaways from our meeting:

DARPA funded future weapons technologies that had the potential to be game changers.

DARPA programs occasionally failed, but that was expected, based on its mission to develop highly advanced weapons technologies.

DARPA had only about two hundred employees. The agency's relatively small size would make it easy to get to know them. The director was right; in a short time, I was routinely interacting with the key DARPA personnel associated with my projects.

I was surprised at DARPA's size, for at the time, I managed almost twice the number of employees at Honeywell. Like many people, my previous knowledge of DARPA came from the movies. In reality, it is simply another government agency with a specific mission to develop future military game-changer technologies. While I cannot go into detail, working with DARPA's personnel became routine, but the projects were extraordinary. My point in relating this story is to clarify the difference between DARPA and the SCO: DARPA focuses on creating breakthrough military technologies for ten, twenty, or even thirty years out, not for the near term; in the common vernacular, we called them blue-sky projects that were "DARPA hard." The SCO focuses on existing needs and applying current technology to address them.

When Secretary of Defense Carter unveiled the existence of the SCO, I had little understanding of the impact it was about to have on the military-industrial complex. Now in the public eye, the SCO and its then director, William Roper, began reaching out to the representatives of industry and the defense community to gather concepts and proposals to help the Pentagon with its near-term requirements.

The existence of the SCO also helped explain Deputy Defense Secretary Bob Work's comment in 2016 that the Third U.S. Offset

Strategy's initial vector was "to exploit all the advances in artificial intelligence and autonomy and insert them into DOD's battle networks to achieve a step increase in performance" that the department believed would "strengthen conventional deterrence."[16] To those of us in the defense industry, we thought it would likely be an SCO initiative. We also knew that we could take existing weapons technologies and propose new applications to the SCO to address near-term needs.

In 2015 the Obama administration created another organization, the Defense Innovation Unit (DIU). Initially, the DIU's mission was to work with Silicon Valley and help the military incorporate rapidly emerging, innovative commercial technologies. The DIU proved extremely useful to the DOD. As a result, it now has branches in Boston, Austin, and Washington DC, as well as a headquarters in Mountain View, California, the heart of Silicon Valley.

Just as it was becoming clear that the Third U.S. Offset Strategy had both a near-term thrust, via the SCO and the DIU, and a long-term thrust, via DARPA, the world changed. To most people's surprise, Donald Trump won the presidency in 2016. Following his inauguration in 2017, there was little mention of the Third U.S. Offset Strategy. The media began to speculate that it was dead.

Is the Third U.S. Offset Strategy Dead?

Most of us would like to think the U.S. military pursues technology on its merit, independent of politics; however, that is not the case. Politics strongly influences the military's technological pursuits. The reason is simple: each president's administration sets national priorities and submits budgets to Congress to support those priorities, and the final budget determines the funded military programs that the Defense Department and the military-industrial complex address. Unlike in the commercial market, you cannot "make the market" for military products by advertising and reducing prices to stimulate sales. Simply put, the DOD budget guides DOD procurements. For example, when President Bill Clinton took office, his administration put new acquisitions on hold for months while his administration set its national defense priorities. If you did not have a DOD contract in place, it was tough to get a new contract for anything not immediately critical to the national defense. At the time, I was responsible for Honeywell's integrated circuits and sensor sales. About half of my sales were to the DOD. The Clinton administration's temporary hold on new procurements turned out to be a nightmare. We had significant layoffs and, in the end, still lost money. Eventually, the DOD funding started to flow again, but missing even a few months of revenue in a large company can be devastating. This example is one of many. Other presidents also want to ensure their administrations set national defense priorities and budget them accordingly.

If you keep in mind the role politics plays in setting national defense priorities, it provides a clue regarding why the phrase "Third U.S. Offset Strategy" fell out of vogue. The Third U.S. Offset Strategy was the brainchild of Obama's Democratic administration. After Trump's Republican administration took office, little mention was made of the Third U.S. Offset Strategy.

Do we still have a Third U.S. Offset Strategy? The answer is yes but with a caveat. Although the Trump administration seldom uses the Obama administration's terminology, the organizations—SCO, DIU, and DARPA—and the pursuit of the technologies and weapons of the U.S. Third Offset Strategy remain in place and a high priority.[17] Before we discuss this in more detail, we should review a critical weakness in the Third U.S. Offset Strategy.

A Weakness in the Third U.S. Offset Strategy

As mentioned in the introduction, the Third U.S. Offset Strategy focused on artificial intelligence, the backbone of autonomous weapons technologies. Unfortunately, this critical technology is also the United States' Achilles' heel. AI technology is a dual-use technology, meaning in addition to its military applications, it has numerous commercial applications; thus, AI technology is commercially available. Also, some AI technologies are coming out of Chinese-owned companies in the United States. According to National Public Radio (NPR),

Instead of buying an existing U.S. business, these Chinese tech giants come to the U.S. and build new companies from the ground up, in what's known as "greenfield" investments . . . [and] hire away a lot of U.S. employees who might otherwise work for American businesses. . . .

This has become a concern in national security circles because the nature of emerging technology is inherently dual-use: The artificial intelligence algorithms that help speed up your smartphone could also be applied to weapons on the battlefield.[18]

Thus, the U.S. government cannot control AI technology, which is a critical pillar of the Third U.S. Offset Strategy. The dual use of this technology and its access by potential adversaries represent a critical weakness in the Third U.S. Offset Strategy. This vulnerability raises an important question: Will the United States be able to maintain its superpower status in the decades to come?

The Future of the United States as a Superpower

Most likely, the United States will maintain its superpower status in the coming decades. While the United States will in all likelihood continue as a superpower, history may repeat itself. Similar to the way the Soviet Union was once considered a superpower, China may also emerge as a superpower. China may be reluctant to take on the same adversarial role the Soviet Union adopted, however, since becoming a hostile U.S. adversary would presumably lead to additional U.S. sanctions and hinder China's economic growth. Russia is unlikely to reemerge as a superpower as its economy is significantly behind those of the United States and China, and it suffers from systemic corruption. Given Russia's legacy nuclear weapons and its ability to use espionage to gain military parity in specific weapons, however, it is likely to remain in its current position as a player on the international stage.

If this analysis is correct, Cold War II—discussed in chapter 1 as an ongoing state of political and military tension between opposing geopolitical power blocs—will become a "Conventional" Cold

War. We can characterize the Conventional Cold War as a time when superpowers continually seek new and more potent conventional weapons.

The Conventional Cold War leaves us with two critical questions:

1. What will deter Russia's further election meddling and aggression to occupy its neighbors, similar to its occupation of territory in Georgia and Ukraine?

2. What will deter China's further intellectual property thefts, cyberattacks, and sovereignty-driven actions in the China Sea?

From my perspective, the answers should not involve open conflict with conventional weapons. The might of conventional weapons is approaching the devastating power of nuclear weapons, so they are likely to have the same dire consequences for humanity. For example, high-intensity laser attacks on an adversary may result in a "laser winter" similar to a nuclear winter, which is a by-product of the smoke rising from the fires following a nuclear attack. The destructive power of high-intensity lasers may result in the same type of fires, thus giving rise to a winter that would be just as dangerous a threat to humanity as a nuclear winter. Either could lead to humanity's extinction.

While humanity has amply displayed its warlike aggressive nature, people are generally not suicidal. We have exhibited restraint in using nuclear weapons; thus, I believe we will likewise show restraint in using those conventional weapons that can replicate nuclear devastation. Given this information, I think the United States, alone, will be unable to deter future Russian and Chinese military aggression.

The critical word in the preceding sentence is "alone." Both Russia and China lack one thing that the United States has—namely, alliances. A recent article in the *Hill* termed them the United States' "Fourth Strategic Offset."[19] This excerpt explains why these alliances are considered a strategic offset:

America's unique and inimitable strategic advantage is our global alliance and partner network. There is none like it, and neither China nor Russia can create a similar network, despite their efforts. This global archipelago of like-minded states includes our treaty allies in NATO, Japan, Korea, Thailand and ANZUS [Australia, New Zealand, and United States Security Treaty], as well as dependable partners such as Sweden, Israel and potentially India, among others. Those who argue geography is irrelevant in the digital age are sadly mistaken. There is no substitute for presence, and our allies and partners allow us a global presence no peer competitor or technological enhancement can match—a presence that acts as a deterrent against aggression of the heavy metal or the gray zone type.[20]

In this Fourth U.S. Offset Strategy, military alliances thus share an equal footing with military technology.

The Fourth U.S. Offset Strategy

Just as each presidential administration sets a national security strategy, the Trump administration did so in establishing the "2018 National Defense Strategy." The respective report recognizes that we Americans "are emerging from a period of strategic atrophy, aware that our competitive military advantage has been eroding." It views China, Russia, North Korea, Iran, and terrorist groups as threats to peace. It sets a new priority in dealing with these threats: "Interstate strategic competition, not terrorism, is now the primary concern in U.S. national security."[21]

Technologies and weapons are highlighted in the first section of the "2018 National Defense Strategy": "New technologies include advanced computing, 'big data' analytics, artificial intelligence, autonomy, robotics, directed energy, hypersonics, and biotechnology— the very technologies that ensure we will be able to fight and win the wars of the future."[22] For the most part, the strategy continues those military technology strategies of the Third U.S. Offset Strategy, but it includes additional strategic thrusts, such as directed-energy weapons and hypersonic missiles, glide vehicles, and projectiles.

The second section of the document speaks to the importance of alliances: "Mutually beneficial alliances and partnerships are crucial to our strategy, providing a durable, asymmetric strategic advantage that no competitor or rival can match. This approach has served the United States well, in peace and war, for the past 75 years. Our allies and partners came to our aid after the terrorist attacks on 9/11 and have contributed to every major U.S.-led military engagement since. Every day, our allies and partners join us in defending freedom, deterring war, and maintaining the rules which underwrite a free and open international order."[23]

While the Trump administration does not explicitly state that the "2018 National Defense Strategy" is the Fourth U.S. Offset Strategy, it is. To my eye, it includes three significant strategic changes to the Third U.S. Offset Strategy:

1. It recognizes the greatest threats to U.S. national security are China and Russia, which have modernized their militaries during the last two decades.

2. It emphasizes new military technology thrusts in directed-energy weapons; hypersonic missiles, glide vehicles, and projectiles; and the militarization of space.

3. It views alliances as an asymmetric strategic advantage.

Various articles online state the Third U.S. Offset Strategy is dead. In a sense, they are correct. The Trump administration rarely mentions it. Its most essential elements, however, are present in the administration's defense strategy, along with the three previously delineated significant strategic changes. In reality, the offset strategy continues but under a different name, the National Defense Strategy. For me, the change brings to mind the old advertising slogan "new and improved."

In part 2, we explore the new directed-energy weapons addressed in the "2018 National Defense Strategy."

TWO

DIRECTED-ENERGY WEAPONS

4

Laser Weapons

New technologies are rapidly giving rise to unprecedented methods
of warfare. Innovations that yesterday were science fiction could
cause catastrophe tomorrow, including nanotechnologies,
combat robots, and laser weapons.

—PETER MAURER

W*ar of the Worlds*, an 1898 science fiction novel by H. G. Wells, introduced lasers to its readers long before their invention. Here is an excerpt:

> It is still a matter of wonder how the Martians are able to slay men so swiftly and so silently. Many think that in some way they are able to generate an intense heat in a chamber of practically absolute non-conductivity. This intense heat they project in a parallel beam against any object they choose . . . it is certain that a beam of heat is the essence of the matter. Heat, and invisible, instead of visible light. Whatever is combustible flashes into flame at its touch, lead runs like water, it softens iron, cracks and melts glass, and, when it falls upon water incontinently that explodes into steam.[1]

Wells awes me with his startlingly close description of a laser weapon's destructive power. This description from *War of the Worlds* may cause you to wonder: What exactly is a laser?

A laser is a device that emits light of identical wavelengths, known as coherent light, in a very narrow beam. The energy of the laser beam can vary widely, from the low power used in laser pointers to the high heat used to cut metal. (To delve into the technical aspects of how a laser operates, please consult appendix B.)

Next we discuss the laser's invention so we have a deeper understanding of it before addressing its applications. Prepare yourself for a historical account with twists and conflicts.

The Invention of the Laser

The question of who invented the laser is controversial. Several scientists worked on the laser's invention in parallel, some even consulting with each other, but then bitterly fought over the patent rights after its invention.

In 1917 Albert Einstein laid the foundations for laser technology. In a paper published that year, he predicted the phenomenon of stimulated emission, which is the process of electrons absorbing energy and then emitting it in the form of light. This phenomenon is fundamental to the function of all lasers.[2]

In 1952, based on Einstein's theory of stimulated emission, three physicists outlined the theoretical principles governing the operation of a maser, an acronym that stands for molecular amplification by stimulated emission of radiation.[3] The maser preceded and laid the foundation for the laser. The critical difference between the two is that a maser emits light with a longer wavelength than a laser does. Since the energy emitted is a function of the wavelength, the maser emits a lower energy light beam than a laser.

In the Soviet Union, Nikolay Basov and Alexander Prokhorov, two physicists from the Lebedev Institute of Physics, presented the theory of the maser at the All-Union Conference on Radio-Spectroscopy held May 1952.[4] In June 1952 American physicist Joseph Weber of the University of Maryland also presented the theory of the maser during the Electron Tube Research Conference in Ottawa, Canada.[5]

Even though the Soviet physicists delineated their theoretical description of the maser a month earlier than the American physicist did, I doubt there was any plagiarism. Having made presentations at conferences, I know the submission of suggested papers to a conference selection committee occurs months in advance. Based on this knowledge, I think the theory of maser development occurred simultaneously and independently in the Soviet Union and the United States.

In 1953 physicists Charles Townes, James Gordon, and Herbert Zeiger built the first maser at Columbia University.[6] In 1964 Townes, Basov, and Prokhorov shared the Nobel Prize in Physics "for fundamental work in the field of quantum electronics, which has led to the construction of oscillators and amplifiers based on the maser—laser principle."[7] Here is where things get interesting, and the story takes its first twist. The American physicist Weber, one of the first to delineate the maser's theory, did not share in the Nobel Prize. After Weber presented his paper in Ottawa, RCA offered Weber a $50 fee, which he accepted, to give a seminar on his theory.[8] Charles Townes learned of it and wrote to Weber, asking him for a copy of the paper, and Weber complied.[9] Townes and his colleagues made a working maser, as did Basov and Prokhorov. Weber's work was theoretical. Although Weber was a nominee for the Nobel Prize in 1962 and 1963, he remains a footnote to its award for the maser. The Nobel Committee considered the maser's "construction" crucial to awarding the prize, even though Einstein won the Nobel Prize in 1921 for his theoretical explanation of the photoelectric effect (i.e., light causing a surface to emit electrons) and never did physical experiments. Somehow, theoretical physics was more important in 1921 than in 1962. In my opinion, Weber deserved to share the Nobel Prize. I think the Nobel Committee did not understand that Weber's paper helped Townes construct the maser.

Are you ready for the next twist? James Gordon and Herbert Zeiger did not share in the Nobel Prize either, even though they had worked with Townes to build the maser. Gordon was a graduate student and Zeiger, a postdoctoral researcher, while Townes was a professor at Columbia University and served as the executive director of the Columbia Radiation Laboratory. The Nobel Committee apparently judged Gordon's and Zeiger's contributions minimal relative to that of Townes.

Does all of this come across as unfair by the Nobel Committee? You be the judge. Meanwhile, prepare yourself for one of the longest patent wars in history as we continue to review the invention of the laser.

The maser has several useful applications, such as enabling atomic clocks and serving as a low-noise microwave amplifier in radio telescopes; however, it lacks the higher energy associated with light of shorter wavelengths that is typical of a laser. When we look at their respective applications, it becomes evident that the laser is more useful for industrial and military uses. Physicists knew about this limitation at the time but had little success in addressing it. The American Institute of Physics, an organization that promotes science and the profession of physics, noted, "[Physicist Charles] Townes thought about the problems intensively. One day in 1957, studying the equations for amplifying radiation, he realized that it would be easier to make it happen with very short waves [higher frequency corresponding to higher energy] than with far-infrared waves [lower frequency]."[10]

It seems the old adage is true: "Great minds think alike." Gordon Gould, a graduate student at Columbia University, after discussions with Townes, built the same type of new maser. As part of his thesis, Gould was working with "pumping" atoms (i.e., adding energy to them) to higher energy states so the laser would emit energetic light. His experiments led him to conclude that he was on to something far beyond the current buzz of masers. In fact, in his notebook, he wrote the word "LASER," which stood for light amplification by stimulated emission of radiation.[11]

Gould, Arthur Schawlow, and Townes—and likely other researchers around the world—now understood how to build a laser. It was only a matter of time before someone would turn the concepts into reality and make a working laser. Gould knew this and feared that others might apply for a patent ahead of him. In April 1959 he filed patent applications with his employer, Technical Research Group (TRG). His concerns turned out to be justified. Nine months earlier, Schawlow and Townes had filed a patent application on behalf of Bell Laboratories. Like steel hitting flint, these opposing patent applications sparked the laser patent wars.[12]

The Laser Patent Wars

After the U.S. Patent and Trademark Office granted Bell Labs a patent on the laser, Gould sued. Using his laboratory notebook, dated back to 1957, he demonstrated that he accurately drew the design of a laser, which, if built, would work. His journal was notarized. On that basis, he argued he was the first to conceptualize how to make the laser.[13]

Another old adage, "The wheels of justice turn slowly," turned out to be true. The legal battles dragged on through the courts for the next thirty years. During that time, the stakes grew higher as the laser industry grew. Finally, in 1987, Gould and his backers started to win settlements, marking the end of one of the most significant patent wars in history.[14] As a side note, if the patent war had ended before 1964, Gould might have shared in the Nobel Prize with Townes, Basov, and Prokhorov.

Who Invented the Laser?

According to the U.S. Patent and Trademark Office, Gordon Gould invented the laser; however, the history surrounding the laser's invention remains unsettled. In a sense, Gould was the first to integrate the numerous concepts in his laboratory notebook, eventually allowing him to prevail in court, but the U.S. court's rulings are still controversial.[15] Meanwhile, many people patented ideas related to the laser, and in addition to Columbia University, TRG, and Bell Labs, many other companies were working on it: Westinghouse Research Laboratories, IBM, and Hughes Laboratories. All had little success, except Hughes Laboratories, which assembled a ruby laser. On May 16, 1960, it emitted pulses of red light, making it the world's first laser.[16] Although Hughes Laboratories did not invent the laser, it was the first to move from concept to reality. At last, with the building of this ruby laser, the world had acquired the ability to direct energy.

I clearly remember the excitement surrounding Hughes Laboratories' achievement. Almost all of us in the scientific community viewed the laser as a remarkable invention, but at the time, we had no idea of its applications. Some called it a solution looking for a problem; however, that view quickly changed. Today, the laser is a pillar of modern society.

Commercial, Industrial, and Medical Laser Applications

The laser has numerous applications. In this section, I only delineate a few in three categories that significantly affect our everyday lives. Doing otherwise would be a book unto itself.

Commercial: If you enjoy watching a movie on a DVD or Blu-ray player or listening to music on a CD player, then you are using a device that has an optical laser in it. The laser scans the data on the disc to pick up the signal. These devices are in most homes throughout the world, making lasers ubiquitous.[17]

Industrial: One of the earliest uses of lasers was in industry. Manufacturers use lasers for cutting, drilling, welding, and engraving, to name a few applications. In addition to enabling high precision, lasers do not induce mechanical stress, as mechanical drills and blades do. If you ever received a plaque as an award, chances are it was laser engraved. If you wear a watch, there is a high probability that the manufacturer used a laser to produce some of the parts. Manufacturers employ a laser for almost every item they produce.[18]

Medical: One of the most widely used medical applications of lasers is in eye surgery, such as in repairing detached retinas and in correcting the vision of people who are nearsighted, are farsighted, or have astigmatism.[19] Another widely used medical application is in medical operations, where surgeons use lasers as a scalpel to cut tissue while minimizing bleeding.[20]

They are only some of the most pervasive commercial, industrial, and medical applications. There are many more, but in the interest of brevity, I am stopping here. As the focus of this chapter is laser weapons, next we cover the military applications of lasers in significant depth.

Lasers struck a harmonious chord in the U.S. military psyche from the first time its leaders learned that a new device might produce scorching beams of light. They dreamed of developing beam weapons capable of replicating those in H. G. Wells's *War of the Worlds*.[21] These dreams would grow and become a reality in the coming decades but not before encountering some significant disappointments.

When Gordon Gould left Columbia University in 1959 to work for the Technical Research Group, the company was entrenched in military projects. Gould got TRG interested in using lasers as weapons, and TRG asked the Department of Defense for $300,000 to research them.[22] According to Science Clarified's *Lasers*, "The military was so fascinated by the idea of beam weapons that it gave the company [TRG] almost a million dollars, more than three times the amount that had been requested. . . . Soon, various branches of the military—army, navy, and air force—gave money to several other companies to develop such weapons."[23]

By 1962 the U.S. military was spending about $50 million a year on laser development. With all this money flowing in, the U.S. military had certain expectations about a laser that would serve as a weapon; however, lasers at the time were unable to emit sufficient energy levels to make a suitable weapon. By 1968 some experts concluded they had hit a wall. The engineering was extremely complicated, forcing them to question if they were trying to violate the laws of physics.[24] I know from experience, though, that engineers are typically optimistic. The hundreds of engineers I managed while serving as a director of engineering for Honeywell usually thought the solution to any problem was just one experiment away. Given money and time, they would grind through experiment after experiment, always believing the next would yield success. Of course, given enough money and time, it usually did. And that is precisely what happened.

In the 1970s the military finally got a weapon—not the *War of the Worlds* weapon it wanted but a close combat laser assault weapon (C-CLAW). It could be used to blind enemy pilots, soldiers, and optical sensors from distances of greater than a mile.[25] Under the direction of the army, its development continued into the 1980s. A 1983 article in the *Washington Post* described its application: "The portable laser beam would sweep back and forth across a battlefield blinding anyone who looked directly at it."[26]

In testing, the C-CLAW could cause the eyes of laboratory animals to explode. The thought of using this on humans, even in war, raised serious ethical questions. Since the army's laser would not discriminate between human eyes and optical sensors, the world community, primarily through the United Nations, considered such blinding lasers horrid. In many ways, blinding lasers were similar to chemical weapons that could cause blindness, and the UN had already banned such weapons. As a rule, the UN seeks to ban weapons that cause "superfluous injuries and/or unnecessary suffering for little military purpose."[27] By 1995, before their use in combat, the UN established the Protocol on Blinding Laser Weapons, Protocol IV to the 1980 Convention on Prohibitions or Restrictions on the Use of Certain Conventional Weapons, that prohibited the use of blinding lasers in warfare and controlled their transfer to any state or non-state actor.[28] The protocol went into force in 1998.[29]

The Protocol on Blinding Laser Weapons has 108 parties (i.e., member states that are signatories to the treaty), including the United States, Russia, and China. Both China and Russia were early signatories, joining in 1998 and 1999, respectively; while the United States did not become a signatory until 2007. Why did it take the United States almost a decade to sign the treaty? I think at the time, 1998 to the mid-2000s, the United States was the leader in this type of laser weapon and was reluctant to limit its options to use it. However, by 2007, other nations likely achieved similar laser capabilities. With the prospect of a potential adversary using a blinding laser on American forces, the leadership decided to become a signatory. If this analysis is correct, it suggests the prospect of having superior conventional weapons trumps ethical concerns. The *Washington Post* reported that the U.S. Army's only concern was "over the probable public reaction once the purpose of the weapon becomes known."[30]

In the 1980s research on directed-energy weapons intensified, as Ronald Reagan's administration established the Strategic Defense Initiative (SDI). The Atomic Heritage Foundation, a nonprofit organization dedicated to the preservation and interpretation of the Manhattan Project and the Atomic Age and its legacy, observed, "Reagan's interest in anti-ballistic missile technology dated back to 1967 when, as governor of California, he paid a visit to physicist Edward Teller at the Lawrence Livermore National Laboratory. Reagan reportedly was very taken by Teller's briefing on directed-energy weapons (DEWs), such as lasers and microwaves. Teller argued that DEWs could potentially defend against a nuclear attack."[31]

When Reagan won the presidency, he wanted a weapon that could shoot down enemy missiles from space. He believed that SDI was a path to ridding the world of nuclear weapons. The Soviets were highly vocal opponents to the initiative, arguing, "SDI would pave the way for weaponizing space."[32] In response, Reagan suggested the United States might eventually share SDI with the Soviet Union, which in an earlier speech, he had described as an "evil empire."[33] Given Reagan's characterization of the USSR, the Soviets distrusted his administration and remained entrenched in their opposition to SDI. Other national leaders worried SDI would set off another arms race and make the Cold War even more dangerous. Nonetheless, despite Soviet protests and shocked officials around the globe, the Reagan administration went forward with the research.

The press dubbed SDI "Star Wars," based on the laser-like weapons used in *Star Wars*, a popular science fiction movie at the time. Much of the program's development focused on two directed-energy weapons—X-ray lasers and subatomic particle beams. (Subatomic particle beam weapons use high-energy subatomic particles to damage a target by disrupting its atomic and molecular structure. Because they do not travel at the speed of light, such weapons are beyond the scope of this book.) Once again, the power needed to supply such weapons proved prohibitive, and SDI floundered.

The U.S. military eventually learned to use low-power lasers to guide missiles to their targets and as range finders on its tanks, as evident in the First Gulf War.[34] By 2006 the U.S. military also used low-power lasers to "dazzle" (i.e., cause temporary blindness) enemy combatants in Iraq. The dazzlers, mounted on the American soldiers' M-4 rifles, provided a nonlethal way to stop drivers who attempted to run through checkpoints.[35]

As is evident in weaponizing the laser, the enormous power requirement proved a tough engineering challenge; however, technology was progressing. The guidance and control systems, built with integrated circuits, were shrinking, becoming more capable, and requiring less power. They were following Moore's law, which is an observation named after Gordon Moore, one of the founders of Intel. Moore's law states the speed and capability of computers will double every two years, as the density of transistors doubles on an integrated circuit during the same period.[36]

Following Moore's law, the guidance and control circuity had improved from 1980 to 2007 by more than eight thousand times. Having worked in the integrated circuit industry, I can attest that Moore's observation was rock solid through 2007. Integrated circuit manufacturers of commercial products, such as solid-state memories and microprocessors, had to plan their future products to fit Moore's law. Failing to do so meant they would not be competitive. In a sense, Moore's law became a self-fulling prophecy. In the last few years, though, some experts in integrated circuits are claiming that Moore's law is reaching an end. The size of various features on advanced integrated circuits have shrunk to only a few atoms wide. On that basis, experts state integrated circuits can no longer possibly follow Moore's law. In one sense, they are correct. In the integrated circuit industry, the ever-shrinking feature size is how companies were able to double the circuit density every two years. However, Moore's observation is a subset of a more encompassing law—namely, the law of accelerating returns. Ray Kurzweil published an essay describing it in 2001, stating, "An analysis of the history of technology shows that technological change is exponential, contrary to the common-sense 'intuitive linear' view."[37]

By using the word "exponential," Kurzweil is referring to how technological change increases in a nonlinear fashion, similar to Moore's law and its prediction that the speed and capability of computers will double every two years. That means such change improves 200 percent (relative to its initial ability). If it were linear, it would only grow 100 percent. I have seen Kurzweil's lecture on the law of accelerating returns. He discusses, for example, vacuum tubes and demonstrates how they evolved exponentially, getting smaller and more capable with time. Eventually, as vacuum tube technology ran its course, integrated circuits picked up where it left off, continuing the evolutionary process and resulting in smaller and more capable electronic systems. My point is, any well-funded area of technology, such as laser technology, will follow the law of accelerating returns and improve exponentially.

What appeared impossible in 1980 thus became routine by 2007. Laser design became a well-known area of expertise, with textbooks explaining methods to overcome hurdles. Technologists also discovered a crucial element to turn a laser into a weapon; —that is, they pulsed the laser to fire in discrete blasts timed in fractions of second intervals. This pulsing causes the target material's surface to explode with each impact, allowing the plasma the laser creates to dissipate. This technique requires less energy for the laser to blast through an object while creating broader craters. After fifty years of funding, the U.S. military was getting closer to what it wanted, a laser weapon along the lines of *Star Wars*. Looking at the evolution of the laser, I conclude it was easier to invent than to weaponize.

In 2014 the U.S. Navy installed the first-ever laser weapon system (LAWS) on the USS *Ponce*, an afloat forward staging base, for field testing. After three months of testing in the Arabian Gulf, the navy reported that LAWS worked perfectly against low-end threats, such as small boats and drones.[38] Following the completion of the tests, the navy authorized the commander of the *Ponce* to use LAWS as a defensive weapon.

LAWS is a 30-kilowatt laser system that looks as if it came out of the 1936 science fiction serial *Flash Gordon* (see figure 1). Until deploying LAWS, most of the lay public had never heard of the USS *Ponce*. Then news of its laser weapon suddenly jolted the *Ponce* into the public eye. The *Ponce*, built in 1966, began to light up the internet as one of the navy's deadliest vessels. With the hype that surrounded the *Ponce* and its laser weapon, as I later wrote in my book *Nanoweapons*, I knew at the time nanotechnology had made LAWS possible.

Nanotechnology is science and engineering conducted at the level of atoms and molecules, typically resulting in feature sizes less than one-thousandth the diameter of a human hair. Without that technology, I doubt the U.S. military and its contractors could have built the laser system. As I wrote in *Nanoweapons*,

> While laser weapons are not a new concept, since their pursuit traces back to the 1950s Cold War, this one ignited the public's imagination, especially Trekkies, familiar with futuristic weapons like phasers. Similar to *Star Trek*'s phasers, the new laser weapon is being refined to provide a wider range of tactical options, such as limiting the damage to a targeted aircraft and stunning an enemy combatant versus destroying the enemy aircraft and killing an enemy combatant. Although the laser weapon's technology is secret, numerous articles argue that the rapid development of nanotechnologies over the last decade enabled significant internal component improvements of the solid-state laser system, making it deployable as a weapon.[39]

Currently, the U.S. military leads in the development of nanoweapons, defined as any weapons that exploit nanotechnology. The U.S. government has been funneling tens of billions of dollars into weaponizing nanotechnology since it established the National Nanotechnology Initiative (NNI) in 2000. The NNI is a research and development initiative involving the nanotechnology-related activities of twenty-five federal agencies.[40]

Based on the law of accelerating returns, it is reasonable to conclude that well-funded nanotechnology, combined with well-funded laser technology, enabled lasers to evolve from dazzlers in 2007 to directed-energy weapons in 2014. Whether you call it a *Flash Gordon* "death ray," a *Star Trek* "phaser," or a *Star Wars* "laser cannon," science fiction became science fact. As noted in chapter 2, similar to the *Star Trek* phaser, the navy's new laser weapon had a dial-in capability. Set to dazzle, it could disable an adversary's drone or boat, allowing its recovery for examination. Alternatively, at full power, it could destroy a drone or boat.

Just as the first detonation of the atom bomb in 1945 ushered in the nuclear age, I think the successful deployment of the laser weapons system in 2014 ushered in a new and potentially more dangerous era, the age of directed-energy weapons. At the time, I remember thinking of the words from the *Rubáiyát of Omar Khayyám*:

> The Moving Finger writes; and, having writ,
> Moves on: nor all thy Piety nor Wit
> Shall lure it back to cancel half a Line,
> Nor all thy Tears wash out a Word of it.

As the first laser pulse left the *Ponce*, traversing space at the speed of light, there was no going back. Warfare changed forever, and our most capable potential adversaries, China and Russia, knew it.

Are High-Energy Lasers Fair-Weather Weapons?

You may have heard the phrase "fair-weather friend," which refers to a person who will stop being your friend in difficult times. Such a person is not a real friend, the type you can count on when needed. The U.S. military has a name for weapons that only work when weather conditions are just right, such as a sunny day—"fair-weather weapons." In general, the U.S. military does not deploy them because it never knows what kind of weather it will encounter during a conflict. Weapons need to work under a broad spectrum of weather conditions, and the U.S. military needs to know it can count on its armaments to work regardless of the elements.

If you have ever used a flashlight in a fog, then you know that foggy conditions interfere with the flashlight's beam. The same holds true for laser beams. Fog and rain, for example, cause problems for commercial lasers, even high-power industrial lasers, by limiting their effectiveness. According to Seyed Hashemipour, Ahmad Salmanogli, and N. Mohammadian, "[The] peak value of the average intensity [of a laser beam] can be astonishingly affected by atmospheric turbulence."[41]

Therefore, the laser has the potential to be a fair-weather weapon. If that were the case regarding LAWS, the U.S. military would not deploy it; consequently, we can reasonably assume the U.S. military has worked to mitigate the effects of weather on its laser weapons. As David Stoudt, writing for Booz Allen Hamilton, the American management and information technology consulting firm, observed, "The HEL [high-energy laser] weapon community has been actively working to mitigate the effects of these conditions for many years, and in the case of atmospheric turbulence, has made significant advances using adaptive optics."[42]

Without going into the details, most of which are classified, the U.S. military's laser can function in multiple atmospheric conditions. This capability has enabled the navy to use the laser in another critical way, as Stoud reports: "The U.S. Navy placed an HEL weapon on the U.S.S. Ponce, where in addition to its capability as a weapon, it was used almost continuously as a . . . reflecting telescope [a telescope that uses mirrors] . . . that allowed visibility of distances greater than 10 kilometers [about six miles] and penetrating things like smoke, haze, and even light fog."[43]

The U.S. Navy is silent on what atmospheric conditions would degrade the effectiveness of the LAWS; however, based on the Booz Allen Hamilton report, it has found solutions to address much of the problem and continues to work on the remainder. In my opinion, LAWS is an all-weather weapon, capable of destroying targets in smoke, haze, rain, and fog. We will only know if I am correct when the U.S. Navy uses LAWS in those conditions and reports it. Short of a conflict, I think the navy will continue to remain silent on this capability.

The Growing Threats

In January 2019, the Pentagon released the "2019 Missile Defense Review," which noted: "In the past several years, for example, North

Korea rapidly advanced and expanded its intercontinental ballistic missile (ICBM) program. Iran extended the range of its ballistic missile systems and may seek to field an operational ICBM. While Russia and China pose separate challenges and are distinct in many ways, both are enhancing their existing offensive missile systems and developing advanced sea- and air-launched cruise missiles as well as hypersonic capabilities."[44]

The threats the United States now faces are more complex than those during the Cold War were. During that period, the United States had one capable adversary, the Soviet Union. We avoided nuclear war because the American and Soviet leadership understood the doctrine of mutually assured destruction. Now the United States faces four capable adversaries. With the rise of new weapons, radical ideologies, and swarming tactics, the doctrine of MAD may not deter nuclear war. Notice in the last sentence that I mentioned "swarming tactics." Let me digress briefly to explain that phrase.

Swarming is a military tactic borrowed from nature. For example, when bees attack, they attack in swarms. In applying this military tactic to sink a U.S. aircraft carrier, an adversary is likely to attempt to overwhelm the defenses of the carrier group by attacking it, for example, with large numbers of missiles. The U.S. Navy relies on the Aegis combat system to enable a carrier strike group to combine powerful computer and radar technologies to track and guide weapons and then destroy enemy targets, such as incoming missiles. (As an engineer, I worked on developing state-of-the-art integrated circuits for the system. The Missile and Surface Radar Division of RCA originally developed the system, but Lockheed Martin currently manufactures it.) The navy has continued to update the Aegis combat system over the decades since its deployment in 1983.[45] In addition, the navies of Australia, Japan, Norway, South Korea, and Spain are using it, with a hundred Aegis-equipped ships now deployed. It also serves as part of NATO's missile defense system.[46] The Aegis combat system is the best missile defense system in existence and can engage medium-range ballistic missiles in flight, but the exact number of ballistic missiles it can intercept is classified.[47] At some point, a large number, or swarm, of incoming ballistic missiles fired at an Aegis-equipped carrier may overwhelm the system's defensive capabilities. That would represent a successful swarm attack.

Given the threats previously discussed, the "2019 Missile Defense Review" states, "In accordance with the FY [Fiscal Year] 2017 NDAA [National Defense Authorization Act], DOD is preparing a strategic roadmap for the development and fielding of directed-energy weapons and key enabling capabilities. When completed, this roadmap will inform high-energy laser investments in the preparation of the President's Budget Request for FY 2020."[48]

The outlined plan makes it clear that high-energy laser weapons are a strategic element of the U.S. national defense strategy. I think the Pentagon realizes that deploying laser weapons on navy destroyers, for example, would be critical to defeating swarm attacks against a U.S. aircraft carrier. It does not take much to connect the dots and come to this conclusion. According to a Lockheed Martin press release from March 1, 2018,

> The U.S. Navy awarded Lockheed Martin . . . a $150 million contract, with options worth up to $942.8 million, for the development, manufacture and delivery of two high power laser weapon systems, . . . [for delivery] by fiscal year 2020. . . .
>
> One unit will be delivered for shipboard integration on an Arleigh Burke–class destroyer [the USS *Preble*], and one unit will be used for land testing at White Sands Missile Range [in New Mexico].[49]

The press release does not specify the power of the lasers, but it is reasonable to judge that they will be as or more potent than the 30-kilowatt laser the U.S. Navy currently deploys. Numerous articles have speculated on their power, ranging from 60 kilowatts to 150 kilowatts. At 150 kilowatts, the laser would be five times more potent than the current 30-kilowatt laser, meaning it can do as much damage as the 30-kilowatt laser in one-fifth the time. This increase in power is crucial because it can destroy a target faster and move on to the next threat if necessary. One thing is clear: the U.S. Navy is serious about deploying lasers and integrating them into the Aegis combat system. The options mentioned in the Lockheed Martin contract suggest that if the first two lasers meet the navy's requirements, it will exercise those options and deploy more lasers on board its ships.

The U.S. Army plans to field an even more powerful laser weapon in the 250- to 300-kilowatt range to protect combat troops against drones, artillery rockets, helicopters, and attack jets. According to *Breaking Defense*, "Less than three months after awarding a $130 million contract [to Lockheed/Dynetics] to build a 100-kilowatt [kW] laser, the Army has decided to skip the 100 kW weapon and go straight for a much more powerful one in the 250–300 kW range. Unlike the original design, the higher power level could potentially shoot down incoming cruise missiles—plugging a glaring gap in US defenses against a Russia, China or Iran."[50] This article suggests the navy's laser weapon will be in roughly the same range, exceeding previous speculations.

A Solution Waiting for a Conflict

Initially, the scientific community viewed the invention of the laser as a solution looking for a problem. As a weapon, it is reasonable to imagine it as a solution waiting for a conflict.

Here are the most compelling reasons why I think the U.S. Navy is eager to deploy lasers on board ships:

Lasers with sufficient power and integrated into the navy's Aegis combat system have the potential to overcome a broad range of threats, from swarm attacks to carrier-killer missiles.

Lasers are cost-effective, with a typical shot to destroy a drone costing less than a dollar in electricity.

Lasers allow unlimited shots, only requiring that the ship's generator pump sufficient power.

Lasers remove a war vessel's major vulnerability by replacing conventional weapons that use gun powder, thus eliminating the need to store ammunition within the ship's magazine.

Lasers make it easier to hit a target since there is no need to calculate the trajectory, the windage, or the target's movement. The laser's trajectory is flat, unaffected by windage, and travels at the speed of light, making even a hypersonic missile look as though it is standing still.

The U.S. Army's reason for building an extremely high-power laser is to address threats from "jet-powered cruise missiles—which fly lower, slower, and with more maneuverability—have proliferated around the world, even to high-end irregular forces like Iran-backed Hezbollah."[51] In addition, the army would also accrue many of the same benefits cited for the navy's laser system.

Laser Weapons of Potential Adversaries

Before we leave this chapter, let us examine the laser weapons of the United States' two most capable potential adversaries—China and Russia. We need to view both as potentially dangerous threats but for different reasons. China has an economy that can support the research and development necessary to develop laser weapons. Additionally, China supports state-sponsored hacking of U.S. intellectual property, including laser weapons technology. We must take Russia seriously because it has nuclear parity with the United States. While Russia's economy is weak, it is still modernizing its military. Like China, the Russians appear adept at hacking. They successfully hacked the Democratic National Committee and Hillary Clinton's campaign during the 2016 U.S. presidential election. Additionally, according to an April 22, 2020, article in the *New York Times*, "A three-year review by the Republican-led Senate Intelligence Committee unanimously found that the intelligence community assessment, pinning blame on Russia and outlining its goals to undercut American democracy, was fundamentally sound and untainted by politics."[52]

China's Laser Weapons

Significant evidence indicates that China is developing laser weapons. According to *Jane's 360*, "Chinese media have reported that a prototype laser weapon is being tested by the People's Liberation Army Navy (PLAN). An article published on 5 April [2019] on the Sina news website contains several screengrabs taken from footage broadcast by China Central Television (CCTV) showing a trainable optical device mounted on a mobile chassis with a large main lens."[53] Unfortunately, the video resolution prohibits reproducing it here; however, China's laser weapon closely resembles the U.S. Navy's LAWS.

Per the promotional video broadcast by the state-run channel CCTV, the transmission shows China's laser weapon in a ground-based, vehicle-mounted application. According to Sina.com, China intends to deploy the weapon both on land and at sea, including aboard its destroyers as an alternative to the short-range surface-to-air missile. This last statement implies it has a range of about three miles.[54] Beyond talking about potential applications, China provides no evidence of the laser's capabilities.

Additional information indicates China is working on laser weapons. In 2017 China released information about a land-mobile laser weapon that successfully destroyed an unmanned aerial vehicle at a range of about a thousand feet. The *Washington Free Beacon* also reported, "China's military is expected to deploy a laser weapon capable of destroying or damaging U.S. military satellites in low earth orbit in the next year [2020], the Pentagon's Defense Intelligence Agency disclosed in a report on space threats."[55]

Evidently China is using espionage to obtain any information it can on the U.S. Navy's developments. The *Maritime Executive*, a source for breaking maritime and marine news, reported, "[The] U.S. Navy has uncovered evidence of widespread and persistent hacking by Chinese actors targeting naval technology. According to a recent internal review ordered by Navy Secretary Richard Spencer, the service's broader R&D ecosystem is 'under cyber siege,' primarily by Chinese hacking teams."[56]

My view is that the Chinese are doing all within their capability to develop laser weapons. Given their tenacity to hack into the most crucial U.S. intelligence information, combined with their government's funding of advanced weapons, weaponizing their lasers is only a matter of time. One thing is certain: if the Chinese can deploy a laser capable of destroying or damaging low-orbit U.S. military satellites in 2020, as they currently predict, then it would signal China's entry into the laser age.

You may wonder why China emphasizes strengthening its military. The U.S. Defense Intelligence Agency's report of January 3, 2019, states, "Chinese leaders characterize China's long-term military modernization program as essential to achieving great power status. Indeed, China is building a robust, lethal force with capabilities spanning the air, maritime, space and information domains, which will enable China to impose its will in the region."[57]

The last three words, "in the region," are particularly significant. I think China's ambition is to dominate the Asia-Pacific region and, more specifically, defend its sovereignty over the China Sea. The latter may eventually lead to a conflict with the United States. A June 26, 2019, article in the *New York Times* reported, "China is an authoritarian nation that most likely seeks to displace American military dominance of the western Pacific. . . . [T]he idea of China as a dangerous juggernaut, more formidable than the Soviet Union, has become increasingly widespread in the [Trump] administration."[58]

Unfortunately, I think it would be difficult for the U.S. military to fight a war with China. China is a nuclear power and may be inclined to use nuclear weapons if its communist-run government thought it might lose such a conflict. If China did resort to launching nuclear weapons, however, it would endanger the survival of all humanity, as we discussed in chapter 2.

I think both the Chinese and American leadership know what is at stake. In 2018 the United States and China were each other's largest trading partners. According to the Office of the United States Trade Representative's website, "U.S. goods and services trade with China totaled an estimated $737.1 billion in 2018. Exports were $179.3 billion; imports were $557.9 billion. The U.S. goods and services trade deficit with China was $378.6 billion in 2018."[59] Therefore, both countries have more to gain by avoiding war than by engaging in one. Additionally, a conflict in the Asia-Pacific region would disrupt trade globally, since roughly one-third of global maritime trade flows through the China Sea.[60]

These salient facts regarding China's laser thrusts, its geopolitical goals, and its trade with the United States place its pursuit of laser weapons in the broader context of its national goal to dominate the Asia-Pacific region. They also delineate the complexities the United States faces as it attempts to confront China.

Russia's Laser Weapons

Russia's interest in laser weapons dates to the Soviet Union's construction of a ground-based laser facility in 1987, a decade ahead of the development of the U.S. ground-based laser at Lawrence Livermore National Laboratory in 1997. On August 19, 1989, the *New York Times* reported, "The department [DOD] has pointed in particular to a facility at Sary Shagan in Kazakhstan, which was alleged to contain a laser weapon that 'could be used in an anti-satellite role today and possibly a ballistic-missile defense role in the future.'"[61]

Upon visiting the facility in 1989 as part of the Soviet glasnost policy of openness, the reporters instead found relatively weak lasers whose "beams were 1,000 times less powerful than those of the Mid-infrared Chemical Laser at the Strategic Defense Initiative's White Sands [Proving Ground in New Mexico]."[62] However, I question if the Soviets intentionally misled the *New York Times*. If the Soviets had revealed a weapons-grade laser at the facility, the *Times* would assuredly have written an article suggesting that the Soviets were capable of attacking U.S. satellites and missiles, information that would have sent shock waves through the Pentagon.

I support my supposition with these facts: An earlier *Times* report stated the Sary Shagan facility was "high atop the region's tallest mountain, is an elaborate complex, bristling with roads, buildings, laboratories, and domes, and linked by heavy power cables to the nearby Nurek hydroelectric plant, one of the largest in the Soviet Union."[63] This description is what I would expect of a laser weapon facility. Furthermore, during the same week as the reporters' visit, the Soviets demonstrated "to American experts . . . a high-power gas laser" at another facility.[64] The demonstration was again part of the Soviet glasnost policy and is proof positive the Soviets had a weapons-grade laser.

These facts raise a question: Were the Soviets being open but misleading at the same time? My answer to the question is yes. My thinking is the Soviets were deceiving the *Times* reporters by showing them only low-power lasers at the Sary Shagan facility but demonstrated a high-power laser at another facility. It reminds me of the rigged three-shell game, where a participant is never able to find which shell hides the pea.

During the late 1980s, the U.S. government was deeply concerned regarding the Soviet's laser capabilities. The *New York Times* reported in 1987 that "Secretary of Defense Caspar W. Weinberger [under President Reagan] recently has warned of powerful new Soviet lasers on the horizon. 'We expect them to test ground-based lasers for defense against ballistic missiles in the next three years.'"[65]

Weinberger's statement is clear evidence the U.S. government worried the "balance of terror" would tilt in favor of the Soviets. When the USSR collapsed, Russia inherited its technology. It makes sense that the Russians would continue to develop the laser technology handed down from the Soviet era. I firmly believe they did. Per the same 1987 *Times* article, "The Most Striking fact about the Soviet Star Wars program is its age and consistency. Anatoly Fedoseyev, a winner of the Lenin Prize and the Hero of Socialist Labor Award for his designs of antimissile radars before he fled the Soviet Union in 1971, observed, 'Since the beginning of Soviet S.D.I., about 35 years ago, this project has never been interrupted or delayed. And I'm sure it never will be.'"[66]

Although the Soviets in the 1990s decommissioned and abandoned the Sary Shagan facility, in 2017 a modernization of the site began. In 2018 Russia tested an anti-ballistic missile interceptor at the site.[67] The days of glasnost are over. Today, secrecy surrounds the Sary Shagan facility, and news regarding the site is sparse.

2. Russia's truck-mounted laser weapon. Russian Ministry of Defense/Social Media.

On March 1, 2018, Russian president Vladimir Putin announced the existence of a new laser weapon during his State of the Nation address: "We have achieved significant progress in laser weapons. It is not just a concept or a plan any more. It is not even in the early production stages. Since last year, our troops have been armed with laser weapons."[68]

Additionally, *Newsweek* reported, "Accompanying Putin's March 1 speech, in which he revealed an array of new and advanced weapons, was a short video showing what appeared to be a truck-mounted laser system [see figure 2]. The ministry entitled the clip 'Combat Laser Complex,' but Putin said at the time that he was not ready to reveal the weapon's name or any other details."[69] In a subsequent meeting, Putin called the laser weapon Peresvet, a name chosen by the Russian public in tribute to "a famous warrior monk who fought at the 1380 Battle of Kulikovo, which ended the Mongol domination of medieval Russia."[70]

The laser is similar in appearance to the U.S. Navy's laser, and that is about all we know. Its capabilities and purpose remain classified. However, according to Putin, such weapons "will determine the combat potential of the Russian army and navy for decades ahead."[71]

Commentary on Chinese and Russian Laser Weapons

The news coming out of China and Russia flows from their state-run media. Both China and Russia typically exaggerate their weapons' capabilities, and both knew the United States was fielding laser weapons with demonstrated lethality. Their respective leaders were under pressure to show their militaries were deploying similar weapons. If they did not have them, I think they would use their state-run media to issue fake news reports. They offered no evidence demonstrating their lasers can down drones, sink small boats, or compensate for atmospheric conditions. If they had that evidence, I think they would flaunt it. That tends to be their style.

However, we should not discount their military capabilities. Both countries are working on developing laser weapons. As previously stated, relative momentum is critical in terms of military capability. I think, for now, the United States has the lead in laser weapons and is maintaining its edge. Unfortunately, the only way to be sure is in the blood-soaked arena of conflict.

Short of conflict, we have evidence that our adversaries are using directed-energy weapons, such as microwave weapons, covertly against the United States. We review this evidence in chapter 5, "Microwave Weapons."

5

Microwave Weapons

The whole secret lies in confusing the enemy,
so that he cannot fathom our real intent.

—SUN TZU

Microwave weapons may sound as if they are new. They are not. During the Cold War, the Americans feared that the Soviets were attempting to use microwave radiation covertly as a means of mind control. U.S. intelligence officials surfaced this concern in 1953 when they detected a low-frequency microwave signal irradiating the U.S. Embassy in Moscow. It continued through the late 1970s and was called the Moscow Signal.[1]

Let us put this in perspective. In broad terms, microwaves are high-energy radio waves that occupy the upper end of the radio wave band. They travel at the speed of light. But the Moscow Signal microwave transmissions were only five microwatts per square centimeter, or about a million times below the power level of microwave ovens to heat food. This measurement was also a thousand times below the federal safety regulation limiting microwave oven leakage; therefore, it appeared insignificant. However, the U.S. intelligence community worried the Soviets knew something about low-frequency microwave radiation that the United States did not, especially as it might relate to influencing the mental state of American diplomats. Covertly, unknown to both the Soviets and the embassy personnel, the State Department and the Defense Advanced Research Projects

Agency launched a program to evaluate its effects.[2] The lure of having a mind control program was irresistible to many sectors in the U.S. government.

Initially, DARPA exposed monkeys to microwave radiation to see if the Moscow Signal would affect their behavior. In 1966 Richard Cesaro, the DARPA official in charge of the project code-named Project Pandora, concluded that the first monkey tested suffered adverse effects regarding its ability to perform tasks.[3] Consequently, Cesaro recommended the Pentagon investigate the potential weapon applications of low-frequency microwaves. In addition, DARPA initiated a new phase of Pandora directed at human testing. At this point, the program received a different code name, Bizarre, and a reclassification as a special access program, necessitating higher safeguards and access restrictions. In typical military vernacular, it became a black program. However, the Bizarre Project and specifically Cesaro's research methodologies were concerning to its scientific oversight committee. Since Cesaro's work was a black program, the experimental procedure and conclusions were not subject to peer review, meaning no scientist in a similar field could evaluate the findings and ensure the quality of the work and the validity of the conclusions. After nearly five years of research and spending millions of dollars to construct a new microwave laboratory, the DARPA oversight committee became concerned about the scientific soundness of the program. Stephen Lukasik, then deputy director of DARPA, asked Sam Koslov, a former DARPA official, to review the project in 1969. Following his review, Koslov was highly critical of the project, and given Koslov's report, DARPA ended its support. We know from declassified information that the Bizarre Project never reached human experimentation.[4] This omission is indeed fortunate since a 2013 article in the *Indian Journal of Biochemistry & Biophysics* reported that "animals [subjected to low-level microwaves] showed significant impairment in cognitive function and increase in oxidative stress [damage to the brain]."[5]

The Soviets did indeed know about the harmful effects of low-frequency microwaves, as the Americans learned after the fact. The

Soviets launched their microwave attack with the hope that embassy personnel would err in their duties, for example, with clerks making mistakes on encrypted messages. Such errors might have allowed Soviet cryptographers to crack American codes.

Did the Soviets succeed? The information remains classified.

You may want to chalk this up to the risk-taking that was typical of the Cold War era. Both the United States and USSR routinely tested each other in numerous ways. One common approach was to violate each other's airspace to see how quickly each side could respond to the intrusions. You would think, however, knowing the dangers of low-frequency microwave exposure on animals would be sufficient to preclude using that tactic in the future. Unfortunately, that was not the case.

On August 11, 2017, the University of Miami received a nerve-racking call from the Trump administration. According to the *New York Times* report, "American diplomats in Havana were getting sick with headaches, dizziness and hearing loss. Washington needed answers."[6] Surprisingly, no one immediately made the connection to the Cold War's Moscow Signal attack. If anyone did, that opinion was kept out of the public spotlight.

After treating the American diplomats, a University of Miami specialist went to Havana to examine other members of the embassy staff. Twenty-four embassy personnel displayed a broad spectrum of symptoms, including headaches, nausea, hearing loss, and cognitive issues. Many asserted they had heard odd sounds before the onset of symptoms. Secretary of State Rex W. Tillerson characterized the illnesses as "health attacks." He added, "We've not been able to determine who's to blame."[7] However, the administration held the Cuban government responsible and expelled twenty-seven Cuban diplomats.

Tillerson's statement suggests the Trump administration suspected foul play. Cuban officials seemed to agree and thought it might be a rogue element of Cuban intelligence intent on ending President Barack Obama's reconciliation efforts. The mystery deepened when Canadian Embassy employees also became ill and began suffering the same symptoms.[8] Things grew even more bizarre (pun intended) in 2018 when similar illnesses emerged at the U.S. Embassy in China.[9]

Initially, based on the Cuban embassy personnel's reporting that they had experienced odd sounds before the onset of their symptoms, the investigators blamed the mysterious illnesses on a sonic weapon. Some termed the laundry list of symptoms the Havana Syndrome. However, by 2018, the *New York Times* reported the most likely cause was microwaves.[10]

Experts now explain the eerie sounds heard by some of the victims as the Frey effect, named for Allan H. Frey, an American scientist who discovered pulsed or modulated microwaves could cause people to perceive they heard sounds.[11] All the pieces fell into place. America's adversaries were using a microwave weapon. Most of the evidence points to Russia as being responsible. As NBC News confirmed, "Intelligence agencies investigating mysterious 'attacks' that led to brain injuries in U.S. personnel in Cuba and China consider Russia to be the main suspect."[12]

The reports' findings align with the same type of weapon used during the Moscow Signal incident. I tend to believe it is true, but the article states, "The evidence is not yet conclusive enough, however, for the U.S. to formally assign blame to Moscow."

I chose to begin this chapter with the Moscow Signal and the Havana Syndrome incidents to illustrate important points: Low-frequency microwaves represent a real type of directed-energy weapon, and adversaries of the United States are using it. To examine microwave weapons, we should start by understanding what microwaves are.

What Are Microwaves?

Microwaves are high-frequency radio waves.

Let us delve deeper into understanding frequency, which by convention is expressed in hertz, or one cycle per second. To understand "cycle per second" as it relates to waves, imagine you are at the beach. Counting the number of waves that reach the shore in a certain period is a measure of their frequency. Ten waves, or cycles, per minute would convert to about two-tenths of a cycle per second (about 0.16 hertz). By comparison, microwaves are expressed as having frequencies between 300 million cycles per second (megahertz) and 300 billion cycles per second (gigahertz), making them high-frequency radio waves.

Given their high frequency, microwaves are relatively short waves; that is, they have short wavelengths. To understand wavelength, let us return to our beach example. As you watch waves from the shore, the distance between the waves as they crest is their wavelength. For example, ten waves cresting per minute would correspond to waves about a hundred feet long (100-foot wavelength). Thus, three hundred megahertz correspond to microwaves with a wavelength of about three feet long. Similarly, three hundred gigahertz correspond to microwaves with a wavelength of about four hundredths of an inch long.

Microwaves have numerous useful civil applications. Let us examine two of them to gain a deeper understanding before we explore their weapon applications.

Radar: Radar (radio detection and ranging) uses microwaves with higher frequencies and shorter wavelengths. Their shorter wavelengths make it possible to transmit them as a beam in a specific direction. They travel in a straight line until reflected by an object they encounter. Directed at a plane, for example, the radar's reflected waves enable the detection of what type of aircraft it is, its direction, and its speed. Using radar allows traffic controllers at airports to direct aircraft traffic.

A short digression might be instructive. Stealth aircraft have a low radar signature, degrading the radar's ability to detect and track the aircraft, and may appear to more closely resemble a bird on a radar screen. Stealth aircraft achieve a low radar signature by reflecting the radar beam in directions that are not detectable or by absorbing a portion of the radar beam.

Microwave Ovens: Most kitchens have a microwave oven. Microwave ovens operate at the frequency of 2,450,000 cycles per second (2,450 megahertz). As most people know, you cannot see microwaves, but you can see your food cooking inside the microwave oven. Microwave ovens work by channeling the microwave beam directly at the food. The molecules composing the food absorb the beam's energy, making the fat and water molecules vibrate. This vibration causes friction, which in turn generates heat and increases the temperature of the food. This increase in temperature cooks the food. You can look inside the microwave because the microwave door contains a plate of glass covered by a metal mesh screen. The screen reflects the microwaves because the mesh holes, which are too small for microwaves to escape, are large enough to allow visible light to pass through and enable you to see what's cooking inside.

Besides these typical microwave applications, there are many more, including industrial applications. However, let us move on to microwave weapons.

Definition of a Microwave Weapon

A microwave weapon is a device that inflicts damage at a target by emitting focused microwaves. The critical word in this definition is "damage." I call your attention to it because while radar and microwave ovens focus microwaves on a target, their intent is not damage.

Critical Attributes of a Microwave Weapon

Microwave weapons, unlike laser weapons, suffer little to no distortion by weather or atmospheric conditions. For example, they can easily penetrate a fog. By contrast, laser weapons find it challenging to penetrate fog.

High-energy microwave weapons have a long reach, typically measured in tens to hundreds of miles. These weapons can damage humans, electronic systems, and fuel. For example, the Havana Syndrome, similar to the Moscow Signal, left some victims with permanent brain damage. Electronic systems exposed to a pulse of high-energy microwaves will suffer catastrophic failure, even if the electronics are off or disconnected from a power source. The microwave pulse induces surge currents in the electronic circuits, causing damage. In the common vernacular, the system will be "fried" regardless of the system's configuration. (*Note*: A system can be hardened against microwaves and other types of radiation, and we explore this point later in this chapter.) High-energy microwaves can also damage the fuel of a missile, truck, or any other platform. The damage results when the microwaves heat the fuel to the point it explodes.

Like lasers weapons, microwave weapons will continue to function as long as they have sufficient power. This ability is a common thread throughout our discussion of directed-energy weapons. Another common point is that directed-energy weapons can replace some conventional weapons that use gunpowder, thus removing the need to supply and store dangerous ammunition for the replaced armament. For example, if a bomb hits the magazine of a warship, the bomb's explosion will trigger the magazine to explode and may sink the ship. Thus, by replacing conventional weapons, directed-energy weapons can significantly improve safety.

U.S. Microwave Weapons

Antipersonnel Microwave Weapons

There are two types of antipersonnel microwave weapons—neurological and biological.

Neurological Microwave Weapons: These weapons attack the human nervous system, typically the brain. At the beginning of this chapter, we covered the egregious effects of the low-frequency microwaves of the Moscow Signal and the Havana Syndrome. Therefore, projecting low-frequency microwaves at humans is, by definition, a neurological microwave weapon. Although it is nonlethal, it can result in permanent brain damage. The United States is silent about deploying or using this type of weapon; however, we know DARPA built one to study its effects on a monkey (in the Pandora Program). Did the United States weaponize it and make others? I leave it to you to formulate your own opinion.

Biological Microwave Weapons: These weapons attack the body in various ways, such as causing skin irritation or the sensation of hearing loud sounds or voices. Let us examine these weapons in detail.

Skin Irritation: The U.S. military has developed and deployed a microwave weapon termed the Active Denial System.[13] According to Phys.org, "A sensation of unbearable, sudden heat seems to come out of nowhere—this wave, a strong electromagnetic beam, is the latest non-lethal weapon unveiled by the U.S. military this week [March 11, 2012]."[14] The military is intentionally not calling this a microwave weapon because it judges the average person will equate this with using a microwave oven. However, I investigated both its frequency and wavelength, and it fits the definition of a microwave weapon. After conducting interviews with U.S. Marine colonel Tracy Taffola, the director of the Joint Non-Lethal Weapons Directorate, and Stephanie Miller, who measured the system's radio frequency bioeffects at the U.S. Air Force Research Laboratory, Phys.org learned the following information:

> The system output frequency is 95,000,000,000 cycles per second (95 gigahertz) and is superficially absorbed by the skin, leading to the target's immediate instinct to flee (hence its name, Area Denial System).
>
> Its reach, or range, is a thousand meters (0.6 miles).
>
> The U.S. military considers the system its safest nonlethal capability, having exposed 1,100 people and resulting in only 2 people suffering injuries that required medical attention to recover fully.
>
> The U.S. military deployed it in Afghanistan in 2010 but did not use it in an operation.[15]

Frey Effect Weapons: These microwave weapons cause people to perceive they hear a sound. In 2003 WaveBand Corporation received a contract from the U.S. Navy to design a microwave weapon for military crowd control.[16] According to *New Scientist*, the project transitioned to Sierra Nevada Corporation in 2008, and its product MEDUSA (Mob Excess Deterrent Using Silent Audio) is a microwave ray gun that causes people to perceive they hear painfully loud booms. *New Scientist* reports, "MEDUSA involves a microwave auditory effect 'loud' enough to cause discomfort or even incapacitation."[17] Unfortunately, much like the Moscow Signal and the Havana Syndrome, some experts suggest MEDUSA may also cause "neural damage." In addition to the victim's appearing to hear noises and voices, the weapon may disrupt a person's balance, cause fevers, and trigger epileptic-type seizures.[18] The U.S. Army, and potentially the U.S. Secret Service, is using MEDUSA or similar technology as well as the technology described in another U.S. Patent.[19]

In 1996 the U.S. Air Force filed a patent for a "method and device for implementing the radio frequency hearing effect." The patent delineated a device that would cause victims to perceive hearing voices. The U.S. Patent and Trade Office granted the patent in 2002.[20] In fundamental terms, this is how it works: The inner ear has sections filled with air and fluid that are vulnerable to microwaves at specific frequencies. The human head acts as an antenna for microwaves. When the head receives those microwave signals, they slightly heat those inner-ear sections, causing them to expand and shift. The human body does not feel the heat or expansions, but the ear records the shifts. The ear's design requires it to interpret the variations as sound, which is a function of the microwave frequency. By modulating the frequencies (i.e., changing the shifts in the inner ear), it is possible to form words. The volume at which the sound is heard is a function of the power of the microwaves.

Unfortunately, a patent has to describe how the technology works in sufficient detail to guarantee the patentee's intellectual property rights, and this one provides insight on how to build this device. Through the Invention Secrecy Act of 1951, the U.S. government can prevent a patent's disclosure. I think this patent presents a possible threat to national security and should fall under the Invention Secrecy Act as the implications of building this microwave weapon are terrifying. This weapon may induce mental illness or cause a person to act irrationally.

Although it is a grim reality, we need to acknowledge these types of microwave weapons exist. Neurological and Frey effect weapons are extremely concerning given that they have the potential to cause brain damage and make mind control feasible. On this basis, I think the United Nations should consider banning their use.

Typically, hearing odd sounds or voices is a sign of mental illness. However, knowing Frey effect microwave weapons exist, a person suffering these sensory effects may not be mentally ill but the victim of a microwave attack. While hearing odd sounds may induce mental illness, hearing voices may cause a person to perform an evil act. Imagine a victim being a member of the U.S. Secret Service who hears an order to assassinate the president. This scenario could be capable of igniting World War III. This last statement may stretch credibility to the limit, but I think it is possible based on my advertising research.

In my latter years with Honeywell, I was responsible for the marketing, advertising, and sales of their integrated circuits and sensors. When I retired, my wife and I opened an internet sales and advertising agency, Del Monte & Associates, Inc. Over the years, the business focus of our company changed. Today our company's revenue is from my royalties, speaking engagements, and consulting based on my books. Given my several decades of advertising experience, I can share some insider information.

Have you noticed that advertisers run the same advertisement multiple times in any media, such as radio, television, and online? They do so for a reason: they are seeking to achieve effective frequency, or the number of times consumers must be exposed to an advertisement to get them to buy a product or to remember a message. Effective frequency influences the consumers to do what the advertiser wants them to do. According to the *Financial Brand*, an online publication delivering ideas and insights for the retail and banking industry, "Marketing experts like to debate the 'right ways' to calculate effective frequency. Some say repeating a message three times will work, while many believe the 'Rule of 7' applies [repeating the message seven times]. There was a study from Microsoft investigating the optimal number of exposures required for audio messages. They concluded between 6 and 20 was best."[21]

Notice that the required repetition of an advertisement is low, twenty times or less, to coax a consumer to take the action the advertiser seeks. Now imagine a microwave weapon that can implant the same message, with a voice repeating it inside your brain continuously thousands of times, day or night, awake or sleeping. The message may be subtle, even unbelievable at first, but through repetition becomes etched consciously or subconsciously in your brain. In time, you might believe it is your own idea.

Another axiom in advertising is that all purchasing decisions are impulsive. While it is probably not 100 percent true, I think advertising is responsible for a significant amount of impulse buying. I believe most of us tune out advertisements, especially when they are repetitious; however, our subconscious minds still record their message. Thus, when we are shopping at the supermarket, for example, we find ourselves reaching for a particular product whose advertisements previously fed into our subconscious. According to the cable news channel CNBC, a survey by Slickdeals.net of two thousand consumers and their impulse buying "shows they make three of those purchases a week, adding up to $450 a month and $5,400 per year."[22]

Advertisers often appeal to our desires or fears. Imagine implanting the fear of an invasion in a national leader's mind to the point he or she becomes irrational. In discussing my concern about Frey effect microwave weapons, you can likely imagine similar scenarios. Typically, only advertising professionals work with effective frequency messaging. Now, though, the U.S. military personnel responsible for Frey effect weapons likely also know about it too.

Again, I think the United Nations should consider banning the use of Frey effect weapons. Currently, no law prohibits their use against enemy combatants or anyone else, which could include you. In addition to controlling your mind, these weapons could leave it permanently damaged.

I provided this insider information because I believe knowledge about microwave auditory effects is a forewarning. Now you know and may support banning Frey effect weapons.

Let us now turn to the next category of microwave weapons.

As previously discussed, microwaves can damage electrical and electronic systems. On May 19, 2019, the *Daily Mail* published an article with the headline "U.S. Air Force Has Deployed 20 Missiles That Could Zap the Military Electronics of North Korea or Iran with Super Powerful Microwaves, Rendering Their Military Capabilities Virtually Useless with NO COLLATERAL DAMAGE."[23] According to the article,

> Known as the Counter-Electronics High Power Microwave Advanced Missile Project (CHAMP), the missiles were built by Boeing's Phantom Works for the U.S. Air Force Research Laboratory.
>
> The microwave weapons can be launched into enemy airspace at low altitude and emit sharp pulses of high power microwave energy that disable electronic devices targeted.
>
> Mary Lou Robinson, the chief of the High Power Microwave Division of the Air Force Research Lab has confirmed to DailyMail.com that the missiles are now operational and ready to take out any target.
>
> While North Korea or Iran may attempt to shield their equipment, U.S. officials doubt that would be effective against CHAMP.
>
> The project has been advancing secretly ever since the Air Force successfully tested a missile equipped with high-powered microwave energy in 2012.[24]

This report would mark a significant milestone in the deployment of microwave weapons, but the *Daily Mail* is usually considered a questionable source. However, there is some evidence that the report may be valid. For one, Ronald Kessler, who wrote the article, is a former *Washington Post* and *Wall Street Journal* investigative reporter. He is also the *New York Times* best-selling author of *The Trump White House: Changing the Rules of the Game* (2018).

Boeing's website also lists a 2012 news release describing the same weapon. Here is an excerpt: "During the test, the CHAMP missile navigated a pre-programmed flight plan and emitted bursts of high-powered energy, effectively knocking out the target's data and electronic subsystems. CHAMP allows for selective high-frequency radio wave strikes against numerous targets during a single mission."[25]

In addition, CNN reported in 2015 that the "Air Force confirms [it has an] electromagnetic pulse weapon. Boeing has developed a weapon that can target and destroy electronic systems in a specific building."[26] In the report, CNN used the phrase "Boeing 'Lights Out' Weapon," which Boeing itself used in a press release that included interviews with Keith Coleman, the CHAMP program manager, and Peter Finlay, the Air Force Research Laboratory's CHAMP lead test engineer.

With CHAMP shrouded in secrecy and the U.S. Air Force silent on its deployment, we must treat the *Daily Mail* story with a skeptical eye. However, if this advanced missile is deployed, it is a superior electronic warfare weapon because it destroys, rather than jams, electronics. Jamming only temporarily affects systems, which can recover when the attack ceases.

If the U.S. military were deploying CHAMP, it would be a game changer, as these cruise missiles can be released and attack an adversary without detection. Since CHAMP is a ground-hugging cruise missile, an adversary likely would not detect it via radar. Without the necessary electronic systems to respond, an adversary would only know that its ability to counterattack was inoperable. It also might not be capable of determining the nation responsible for its mysterious power loss. CHAMP's pulse would render an adversary's command center useless, with its computers fried, communications salient, and lights out. Thus, the system's capability has the potential of rendering the MAD doctrine void.

Drone Defense Microwave Weapon

Swarming tactics are a reality, and potential U.S. adversaries are using them. On July 22, 2019, Iran seized two merchant vessels—a British oil tanker and an unidentified foreign oil tanker.—by using swarm tactics.[27] According to Reuters, "Instead of trying to match the U.S. military weapon-for-weapon, Iran deploys large numbers of relatively unsophisticated systems on land, at sea and in the air. The idea is to overwhelm American forces, much in the way a single bee is a nuisance to a human being but a swarm of them could prove lethal."[28]

In 2002 the U.S. military launched the war game Millennium Challenge, the most extensive simulation ever held, involving 13,500 people. It ran from July 24 to August 15 and included both live exercises and computer simulations. Its purpose was to simulate a war with Iran set in 2007. According to the *New York Times*, "The upshot was that the enemy 'sank' much of the American fleet as the exercise opened."[29]

Given the might and sophistication of the U.S. Navy, it is reasonable to question how this is possible. The answer is one word, swarming. In the war game, Iranian forces deployed swarms of speedboats armed with cruise missiles, rockets, torpedoes, sea mines, machine guns, and shoulder-fired surface-to-air missiles. In addition, the Iranians deployed shore-based missiles also capable of swarming the U.S. fleet. To the surprise of the war game's participants, the swarming was effective and inflicted significant damage to U.S. Navy warships.

The U.S. military uses these war games as a method to test equipment and concepts. I think lessons learned from the Millennium Challenge sharpened the navy's need for littoral (i.e., close to shore) combat ships to address swarming threats. The navy is currently pondering how to make these vessels more lethal.[30] In addition to conventional armaments, it appears to be leaning toward directed-energy weapons.[31] While the navy is finding lasers are effective against speedboats, Raytheon's advanced high-power microwave system is proving itself a more effective drone killer. In 2018, according to the company's website, "Raytheon's high-power microwave system engaged multiple UAV [unmanned aerial vehicle] swarms, downing 33 drones, two and three at a time."[32]

In my opinion, the high-power microwave system is more effective against drones. The microwave beam disrupts the drone's guidance system and can attack the entire swarm, downing multiple drones at a time. In the same test, Raytheon's high-energy laser system proved lethal against drones but zapped them one at a time.[33]

While the navy is still testing to determine how it will arm its littoral combat ships, directed-energy weapons appear to be in the running. For example, in 2020 the navy stated it would begin testing the effectiveness of laser weapons aboard them.[34]

Comments on U.S. Microwave Weapons

In concluding our short survey of some of the U.S. military's microwave weapons, there are others, but I chose these based on their maturity and because they are representative of the class. To my mind, microwave weapons offer a broader spectrum of capabilities than lasers do.

Two points were touched on that bear repeating:

1. Microwave weapons are far less sensitive to atmospheric disturbances than lasers, making them a more robust all-weather weapon.

2. Microwave weapons appear better suited than lasers against drone swam attacks.

I expect future warships of the U.S. Navy to include both laser and microwave weapons. In combination, they would remove the need for the Phalanx machine gun—the navy's close-in weapon system that serves as a last-ditch defense against missiles and uses gunpowder, a potential liability—and the use of short- to intermediate-range missiles against drones and missiles. Laser and microwave weapons also provide a low-cost, unlimited, and continual defense against missile, drone, and speedboat swarms; short- to intermediate-range missiles do not. The most crucial phrase in the last sentence is "low-cost, unlimited, and continual." As long as the navy supplies power to these directed-energy weapons, they will continue to work, and a typical laser shot costs about a dollar, as mentioned in chapter 4. By contrast, short- to intermediate-range missiles are expensive, typically costing hundreds of thousands of dollars, and a warship can only carry a finite number of them.

Let us now examine the microwave weapons of two of the United States' potential adversaries—Russia and China.

As noted, it is highly probable that Russia has developed a low-frequency microwave weapon. It is also likely the Russians used it against U.S. Embassy staff in Moscow (1953), Cuba (2017), and China (2018). They do not claim to have such a weapon, but significant evidence says they do.

In 2009 Russia and Cuba signed a strategic partnership alliance aimed at expanding mutual cooperation in agriculture, manufacturing, science, and tourism.[35] While there were no public statements regarding their rekindling of Cold War–era military ties, Russia needed military allies, and Cuba needed financial help. Cuba is also conveniently located only about a hundred miles from Florida. These points suggest Russia and Cuba would secretly engage in a military alliance as well. As noted previously, the Cuban government, armed with a Russian microwave weapon, possibly perpetrated the attack on the U.S. Embassy personnel in Havana. I think it is unlikely the Russians would have executed an attack on Cuban soil without at least having the Cubans' approval. My conclusion is that whether the Cubans attacked the embassy directly or allowed the Russians to conduct the attack, the Cubans were involved. This same line of reasoning applies to the attack on U.S. Embassy personnel in China. In my opinion, no military operation would take place in China without the approval of its government. I think China and Russia may be equally responsible for the attack, with the same goals as the Moscow Signal incident. This segue leads to a crucial observation: Russia's ties to Cuba and China may have enabled it to trade this microwave weapon for the secret information about the United States that was gained through the weapon's use.

Russia is aware that the United States is developing microwave weapons. The Russian economy and its corrupt government may hamper its indigenous development of high-power microwave weapons. However, through either espionage or its relationship with the Chinese, whose history demonstrates an ability to hack U.S. military secrets, I believe the Russians obtained the blueprints to build high-power microwave weapons. As I have mentioned before, keeping military secrets over a protracted period is almost impossible. Therefore, it should come as no surprise that in 2015 the Futurism website, which covers the future of science and technology, published the following: "Russia has just announced their creation of a microwave gun with the capability to knock drones and warhead missiles out of the air from 10 kilometers [about six miles] away!"[36]

In typical Russian fashion, the details of the weapon remain secret. Officials reportedly scheduled a private demonstration of the weapon during the Russian Defense Ministry's Army-2015 expo.

If Russia's claims are valid, according to *Military & Aerospace Electronics*, it could "complicate U.S. military strategic planning, which for the past quarter-century has relied heavily on precision-guided munitions, GPS navigation, and tactical battlefield networking."[37] While Russia tends to exaggerate its new weapons' capabilities, this report is four years old. In that time, even with typical development issues, Russia could have engineered it to be a potent microwave weapon.

According to a 2010 research report by Robert J. Capozzella, "As for the anti-aircraft systems, Russia is researching and trying to sell the Ranets-E and Rosa-E. . . . The first is a point defense system designed to target the electronics of modern aircraft; the second is an aircraft defensive system that targets the radar of enemy aircraft. These are still in development, however based on the advertised beam output; [*sic*] their range is promising against unshielded systems but otherwise limited."[38]

Apparently, as part of the sale, Russia requires additional development investment from the buyer, but clearly Russian military leadership intends to build and sell microwave weapons. Capozzella's report is more than a decade old, as of this writing. Meanwhile, the Russians may have secured the necessary development funding and perfected these weapons. Unfortunately, Russia's "iron curtain" still hides secrets from the free world.

My concern is that the relative momentum of U.S. leadership in directed-energy weapons will likely be short. Recall that its leadership momentum in nuclear weapons lasted only four years against the Soviet Union, which detonated its first atomic weapon in 1949.

Chinese Microwave Weapons

China is actively building microwave weapons that appear identical to the U.S. military's CHAMP and Active Denial System. Rather than keeping these weapons secret, the Chinese are touting their capabilities. Let us examine these weapons.

China's CHAMP-like Microwave Weapon: According to a *Popular Science* report in 2017, "For over 6 years, Huang Wenhua and his team at the Northwest Institute of Nuclear Technology in Xi'an have been working on a potent microwave weapon. This one, which recently won China's National Science and Technology Progress Award, is small enough to fit on a lab work bench, making it theoretically portable enough for land vehicles and aircraft."[39]

I do not doubt that China will weaponize the technology to gain a CHAMP-like microwave weapon. I base this conviction on a 2017 report by Richard D. Fisher Jr., who found "it appears increasingly likely that any period of [the United States'] advantage from these weapons could be shorter than expected due to China's large investments in energy weapons development."[40] China is thus strategically making significant investments in directed-energy weapons. The belief by China's leadership that directed-energy weapons will dominate warfare by midcentury is fueling this level of investment.[41]

China's Microwave Active Denial System: This system appears identical to the U.S. Active Denial System, even sharing its name. According to the state-run Chinese tabloid *Global Times*, "China is developing a non-lethal weapon system based on microwave radar technology, which the chief engineer of the project said improves the country's counter-terrorist and land and maritime border defense capabilities.... [It is] officially named Microwave Active Denial System, works by shooting millimeter microwaves at targets, which can cause the pain nerve under the skin to ache in a bid to effectively halt the objective's [a person's] violent actions and disperse targets [crowds]."[42]

Finally, from the attack on U.S. Embassy personnel in China, I think it is evident the Chinese likely have low-frequency microwave weapons, including Frey effect weapons. Again, if the Russians were involved in the attack, they had to have obtained China's approval. I do not think China would provide its consent without getting access to Russia's microwave weapon technology.

This section completes our survey of microwave weapons. Next, I think it is appropriate to discuss how nations are shielding their military hardware against the use of such weapons. I spent a significant portion of my career in this area. Much of my work remains classified, but I can share what is in the public domain.

Radiation-Hardened Electronics and System Shielding

Nuclear weapons can create high radiation environments. Those environments give rise to electromagnetic radiation, which in turn can fry electronics. Since U.S. military hardware has to work in high radiation environments, such as in outer space or during a nuclear detonation, the electronics must be hardened and the cables shielded. However, let me make an important distinction: nuclear weapons create electromagnetic pulses, not the sustained radiation found naturally in outer space or produced by a microwave weapon; therefore, the shielding and hardening required will depend on the equipment's application.

The phrase "radiation-hardened systems" typically means the entire system is radiation resistant. This level of radiation resistance requires the electronics to be radiation tolerant or shielded, and that shielding must also protect the interconnections. If any portion of the system is vulnerable, it may lead to a catastrophic failure. For example, during a microwave or EMP event, even one interconnection without shielding can send high electrical surges throughout the entire system.

Radiation-hardened electronics are components whose design and manufacture allow them to withstand high radiation environments, such as those found in outer space or nuclear environments. I worked on manufacturing such components, and they are in many current military satellites and weapons. Their designs and construction are classified; however, I can attest from experience that these integrated circuits are incredibly robust in radiation environments. Again, though, the levels of radiation and the duration of exposure to which they were exposed are classified.

Honeywell's Solid State Electronics Center was one of the few integrated circuit foundries in the United States capable of producing radiation-hardened electronics. As a director working at the center, I was responsible for the program management of hundreds of engineers, technicians, and support personnel involved in their manufacture. These integrated circuits are challenging to design and produce. Their cost reflects that difficulty.

My colleagues and I were aware that our technology was a target of potential U.S. adversaries. Working on these programs required a secret clearance. I will not go beyond this general discussion because of the sensitive nature of the material. Let me add, though, I believe the United States leads the world in radiation-hardened electronics.

Another essential way to protect electronics and interconnections is by shielding, which typically involves using metals to form a Faraday cage, or a continuous metal enclosure. As noted in chapter 1, a Faraday cage makes it difficult, but not impossible, for a microwave or EMP to penetrate its interior.[43] Therefore, putting your smartphone in a metal can will provide some shielding from an EMP. How militaries shield their systems is, in practice, complex. Various materials can absorb radiation or reflect it. Military system designers also have to balance weight restrictions versus radiation protection. For example, while lead is generally an excellent radiation shield, its weight makes it challenging to use for space applications. Launching heavy satellites into space is extremely difficult.

I recognize that this section provides only a top-level overview. Unfortunately, much of the crucial information is classified, and what is not classified is available from other sources. In appendix C, I have listed several sources for those who wish to delve deeper into the subject. My intent in this section is to demonstrate that countermeasures against radiation environments exist, and to my knowledge, the U.S. military is a leader in these countermeasures.

If you found microwave weapons concerning, the next directed-energy weapon, electromagnetic pulse weapons, are even more so. The reason I find them more concerning is that even weak nuclear powers such as North Korea can launch an EMP attack, which some claim could cause the deaths of up to 90 percent of the U.S. population. This type of attack makes chapter 6 outright scary.

6

EMP Weapons

An EMP attack on America would send us back to the horse
and buggy era—without the horse and buggy.

—TRENT TRANKS

Y our day may start like any other. Like most people, your alarm
may wake you at 7:00 a.m. After going through your morning
routine, you may eat breakfast, head to work, run errands, do
housework, and care for the kids. About noon, you may take a break
for lunch and then get back at it. Around 5:00 p.m., you may break
for dinner. By 10:00 p.m., you may start winding down, carving out
time for personal care, and eventually go to sleep. Does all this sound
familiar? It may since it describes the lives of many Americans.[1]

One day, your routine may change without warning. John Len-
non's line in his song "Beautiful Boy" says it all: "Life is what happens to you while you're busy making other plans." That day, time
will stop, all battery and electrically operated clocks will freeze, and
the lights will go out. Not just your lights but also your neighbors'
lights. Then you'll discover it's not just the neighbors' lights that are
out but all the lights in the city as well. All cities in the state and all
states in the United States will simultaneously darken. What is happening? An adversary has attacked the United States with an electromagnetic pulse weapon, and we are at war.

Your first instinct may be to dismiss the possibility of that ever
happening. You may not have previously given it this much thought.

111

Most Americans do not contemplate an EMP attack on the nation's electric grid, and if they do, they dismiss it as a remote possibility. However, potential adversaries such as Russia, China, North Korea, and Iran are perfecting this warfare strategy.[2] In my opinion, it is one of the most severe attacks that even a weak nuclear power such as North Korea can launch.

I recognize we may be getting a little ahead of ourselves, so let us begin by understanding what constitutes an electromagnetic pulse.

Electromagnetic Pulse

An EMP is an intense burst of energy made up of electrical or magnetic waves. Magnetic waves are invisible, as you probably know if you ever played with magnets. The electric waves are invisible also but just as real.

An EMP can induce pulse currents in unshielded electrical and electronic systems, such as computers, radios, and electric generators. If the EMP is massive, it can fry that equipment, including the generators that support the U.S. electrical grids. This ability to render an enemy's electronics and electrical grid useless makes EMP weapons terrifying. At this point, you may wonder, How are EMPs generated?

Generating an EMP

An EMP can originate in three ways: natural causes, civil causes, and military causes.

Natural Causes: Lightning is a common natural cause of an EMP.[3] For example, a lightning strike on a power line will send a massive surge current into a home's electrical panel. Without a surge protector, the current will cause many of the electronic and electrical systems to burn out. Let me share a personal experience of this type of EMP.

My wife and I raised our two sons in Minnesota, where we still live. In the early 1980s when our sons were in their preteens, a lightning snowstorm knocked out the power for five days. Unfortunately, the storm occurred in the winter, with temperatures hovering in the single digits. Temperatures this cold are life-threatening. The *New*

York Times reported, "[The] Centers for Disease Control and Prevention, which uses methods most in accordance with global standards, currently [states], cold weather kills more people than hot weather does."[4]

Many of our neighbors decided to abandon their homes and take residence in a motel, in a hotel, or with friends or family not affected by the power loss until the restoration of power. My wife and I decided that was a poor strategy, as the outage could result in looting or our water pipes freezing and breaking. We chose to stay in our home to weather the storm.

Fortunately, we had two fireplaces, one in the living room and one in the rec room. I also purchased a kerosene heater and two five-gallon jugs of kerosene. Equally fortunate, we had a cord of firewood, and our kitchen stove used gas, not electricity. The stovetop worked, but the oven did not because it used heat sensors controlled by electricity.

Between the two fireplaces and the kerosene heater, we were able to keep the temperature in the house well above freezing. As a result, none of our pipes froze or broke.

My wife cooked all our meals on the stovetop, and we ate in the kitchen. Unfortunately, we could not keep the gas burners continually lit because they consume oxygen and emit carbon dioxide and water vapor. Eventually, the oxygen level would begin to drop in the kitchen, and the flames would consume carbon dioxide. At that point, they would start generating carbon monoxide, a deadly gas. We noticed this because we had carbon monoxide detectors.

In the evening, my wife and I would play recorded radio programs for our sons and play games with them. Our sons loved the experience and could not wait for another power loss. They can look back at it fondly. My wife and I knew we were dealing with a potentially life-threating situation.

Since the storm only affected the upper Midwest, linemen from other parts of the country traveled to the affected area to help restore the numerous power outages throughout the area. The experience was sobering for all of us and gave us a small taste of how people lived before 1878, the year Thomas Edison filed his first patent application for "Improvement in Electric Lights." (As a side note, Edison did not invent the first electric light bulb. Other scientists created light bulbs before Edison was born, but they burned out in a few minutes. Edison invented the first practical light bulb, hence the word "improvement" in his patent.)

Lightning is the most common natural cause of an EMP. Other natural events, such as a solar flare, can also cause a vast EMP. These flares are rare, occurring about once every hundred years.

Civil Causes: Power surges are the second-most common causes of an EMP. Power surges can occur when the electric utility company performs power grid switching and when large appliances such as air conditioners and refrigerators turn on and off.[5]

To understand the first cause, we need to examine how a power company operates. According to the National Electrical Manufacturers Association Surge Protection Institute, a not-for-profit trade association, "During the operation of the electrical grid, the utility may need to switch the supply of power to another source or temporarily interrupt the flow of power to its customers to aid in clearing a fault from the system. This is often the case in the event of [a] fallen tree limb or small animal causing a fault on the line. These interruptions of power cause surges when the power is disconnected and then reconnected to the customer loads."[6]

You may have noticed during a storm that your lights flicker, which could be the result of the power utility changing power sources and causing a small EMP. To determine why switching the power causes an EMP, let us begin by understanding that electrical current creates a magnetic field. For example, a junkyard uses an electric current magnet to move scrap iron with a crane. The opposite is also true, as Michael Faraday demonstrated in 1831: a magnet moving relative to a conductor produces current.[7] His discovery of this phenomenon is termed electromagnetic induction.

With this understanding, imagine a power company switching from one power source to another. During that process, it has to turn the power off and turn it back on again. This switching causes a profound and sudden change in the magnetic field that, by defini-tion, is an EMP. This EMP produces a power surge, which transmits the EMP to your home. You will not feel the EMP, but your flicker-ing lights are a sign that one is present. It is equally probable that you will not see any effect during power grid switching if the EMP is small. However, even those tiny utility electromagnetic pulses can degrade electronics and appliances over time. Rarely, the EMP transmitted to your home when a utility company performs a power grid switch can exceed the power tolerance of various devices and cause them to fail. An example is a light that brightly flashes and then burns out, similar to an old-fashioned flashbulb.

My wife and I can personally attest to this type of EMP's damage, as our home was the victim of such an event about a decade ago. It fried our TV sets, radios, garage door opener, and refrigerator. My wife was in her studio and told me that her drafting table light went off like a flashbulb. Our home filled with smoke as electrical cords to various appliances and surge protectors melted. Flames shot out of one electrical outlet, destroying it. We called the fire department. When personnel arrived, there were no visible flames, just smoke. They checked the house and walls for any sign of fire. Fortunately, there was none.

Damage to our home and possessions was in the thousands, which our homeowner's insurance policy covered. The insurance inspector told us that a power surge could weaken electrical appliances and cause them to fail during the next two years; therefore, the insur-ance claim remained open for that period. Fortunately, we did not experience additional appliance failures. The power surge also fried the power company's transformer, so we were without power for days until its repair.

The power surge not only affected our home but that of our neigh-bors. Fortunately, the outage that affected us occurred during the late summer, and temperatures were moderate. If it had happened during the peak of summer, life-threatening heat waves might have engulfed us. As the *New York Times* reports, "[The] number of deaths associated with extreme heat . . . increase. And those deaths, according to the National Climate Assessment, will exceed the decline in deaths from extreme cold, meaning an overall increase in mortality."[8] Heat waves can be just as deadly as extreme cold. Any power loss during weather extremes adds to the lethality of an EMP weapon.

As noted, an electromagnetic pulse also can occur when appliances, such as air conditioners and refrigerators, turn on and off. The phenomenon that causes this EMP is the disruption of power that, in turn, disrupts the magnetic fields. However, the EMPs generated by these types of machines are typically small and do no damage to other electronic systems in your home.[9] Sometimes, though, they can cause flickering lights and electric anomalies.

Military Causes: EMP weapons work by causing a short, intense burst of electromagnetic energy. There are two types of EMP weapons. As discussed in chapter 1's scenario, nuclear weapons can create a large nuclear electromagnetic pulse capable of affecting a significant area and are termed nuclear EMP weapons. The second type of EMP weapon is a nonnuclear EMP weapon. In the next sections, we discuss these EMP weapons thoroughly.

Nuclear EMP Weapons

Nuclear weapons can create a massive EMP. These types of EMPs are typically the result of a high-altitude detonation. It turns out that altitude plays a crucial role regarding the power of an EMP and the area affected.

The first test of an atom bomb, Trinity, took place in the New Mexico desert on July 16, 1945, and the U.S. Army shielded the electronic equipment intended to measure its effect. Enrico Fermi, an Italian and naturalized American physicist credited with creating the first atom bomb, predicted the generation of an electromagnetic pulse in concert with the bomb's detonation. According Robert Wilson in the *Trinity* report, Fermi "calculated that the ensuing removal of the natural electrical potential gradient in the atmosphere will be equivalent to a large bolt of lightning striking that vicinity."[10]

In simple terms, Fermi determined that the detonation would produce an EMP equivalent to a lightning strike. In a recurring theme concerning nuclear weapons, the scientists responsible for their development typically underestimated the destructive power of these various weapons. While Fermi predicted a lightning strike–like effect, the EMP turned out to be significantly stronger. As Wilson states, "All signal lines were completely shielded, in many cases doubly shielded. In spite of this many records were lost because of spurious pickup at the time of the explosion that paralyzed the recording equipment."[11]

Based on Fermi's predictions, the U.S. Army attempted to shield the signal lines but failed. The U.S. military called this unique feature of nuclear weapons a radio flash. The term stemmed from the click personnel heard on their radios each time a nuclear device detonated. However, they were a long way from understanding it.

In 1958 the U.S. military got its first clue that altitude played an essential role in the propagation of EMPs. During the detonation of a helium balloon–lofted 1.7-kiloton atom bomb in the Yucca Flat, a nuclear test region within the Nevada Test Site, the EMP was at least five times greater than expected.[12] Another surprising thing happened. Instead of being vertical, affecting a narrow area, the EMP was horizontal, affecting a broad area. Unfortunately, the U.S. military did not place much emphasis on this vital piece of information, mainly dismissing it as a wave propagation anomaly.[13] I say "unfortunately" because their next significant high-altitude test caused much destruction.

On July 9, 1962, the United States carried out a high-altitude test, Starfish Prime, by detonating a 1.44-megaton bomb (approximately a hundred times more potent than the atomic bomb dropped on Hiroshima). The explosion occurred 250 miles above the Johnston Atoll in the Pacific Ocean.[14] It was the United States' most significant nuclear test in outer space. Typically, the U.S. government would have kept this nuclear test top secret; however, once again, the scientists involved miscalculated the EMP effects this device would produce. Keeping it hidden in any sense was impossible.

Nine hundred miles from the space detonation are the beautiful islands of Hawaii. At approximately 11:00 p.m. Hawaii time, the

Honolulu islanders casually gazing at the horizon in the direction of the test saw what they described as a beautiful aurora borealis. The explosion peaked ten degrees over the faint night horizon from the standpoint of Honolulu, and its artificial aurora borealis lasted seven minutes. It was visible across the Pacific Ocean from Hawaii to New Zealand. The beauty of the aurora borealis on that fateful night proved how deadly and destructive an EMP could be.

The detonation took place in the Van Allen radiation belts, which consist of energetic particles that shield the earth from the solar winds (i.e., the continuous flow of charged particles from the sun). James A. Van Allen had discovered them in 1958.[15] At the time of the Starfish Prime test, the crucial role the Van Allen belts played in protecting Earth's inhabitants from harmful solar radiation was unknown. Without them, however, humans on Earth would perish.

The U.S. military wanted to study how nuclear explosions could disrupt the Van Allen radiation belts, and Van Allen, a University of Iowa physicist, agreed to cooperate with the military in this study. According to *Wired*, the U.S. military also had a covert objective: "[The] Starfish Prime was intended to test whether nuclear explosions in low-Earth orbit (LEO) could augment and expand the Earth-girdling Van Allen radiation belts to create a barrier that would incapacitate Soviet intercontinental missiles launched against the United States."[16]

The Starfish Prime test temporarily altered the shape and intensity of the Van Allen belt, creating the artificial aurora borealis. As a result, the explosion pumped additional high-energy particles into the belts and launched an EMP that was literally off the charts. Suddenly, according to *Discover* magazine, in Hawaii "it blew out hundreds of streetlights, and caused widespread telephone outages. Other effects included electrical surges on airplanes and radio blackouts." The detonation also led to the loss of satellites. According to *Discover*, "The pulse of electrons from the Starfish Prime detonation damaged at least six satellites (including one Soviet bird [satellite]), all of which eventually failed due to the blast. Other satellite failures at the time may be linked to the explosion as well."[17]

The last sentence says to me that we do not know the full extent of the damage that the Starfish Prime test caused. Lost in the details are people who died on life support equipment that failed, in traffic accidents due to light signals not working, and from being unable to phone for urgently needed medical attention.

Primarily, the Starfish Prime test's location was arbitrary. According to ThoughtCo., a premier reference site, "Were a nuclear device to be exploded in space over the middle of a continent [the United States, for example], the damage from the EMP could affect the entire continent. . . . A modern EMP from a space nuclear explosion poses a significant risk to modern infrastructure and to satellites and space craft in low Earth orbit."[18]

Although political pundits speak of military intelligence as an oxymoron, the Starfish Prime test finally made it clear to the U.S. military that a nuclear detonation in low-Earth orbit is an EMP weapon. It did not require a nuclear physicist to imagine that one nuclear weapon in the megaton range detonated in a similar orbit over the central United States would leave the country in darkness. However, this test occurred during the height of the Cold War, and the U.S. military did not share this lesson with the Soviets.

Unfortunately, the Soviets did their first significant high-altitude nuclear EMP test (part of the K Project) west of Jezkazgan (located in central Kazakhstan), on October 22, 1962, during the Cuban missile crisis. In this test, the Soviets detonated a 300-kiloton nuclear bomb at an altitude of 180 miles (about five times smaller than the U.S. Starfish Prime test and at about two-thirds the elevation). The Soviets revealed the level of damage they suffered to American scientists in 1995: a fire that destroyed the Karaganda power plant and currents that shut down 620 miles of power cables between Astana and Almaty.[19]

The Starfish Prime and K Project results alarmed both the Americans and the Soviets, who were also nearing a nuclear conflict over the Cuban Missile Crisis.[20] Let me place the era in context for young readers who did not live through it. On October 14, 1962, information collected by an American U-2 spy plane revealed the Soviets were building a ballistic missile base in Cuba capable of attacking points in the Western Hemisphere with nuclear weapons.[21] The prospect of a nuclear war seemed a near certainty during President John F. Kennedy's televised address to the American people on October 22, 1962. In his historic speech, lasting seventeen minutes and thirty-eight seconds, he outlined seven steps to ensure the removal of the offensive ballistic missiles from Cuba.[22]

At the time, I was a teenager living with my parents in Jersey City, just two miles from New York City. I can remember a feeling of dread that nuclear war would ensue and that New York City was a prime target, meaning my home was close to ground zero.

For thirteen days, the possibility of a nuclear conflict between the Soviet Union and the United States was uppermost in the minds of people around the world. Then almost as suddenly as it started, calmer minds prevailed, and the Cuban Missile Crisis ended without a single shot fired. The superpowers reached a deal: the Soviets would dismantle the Cuban missile sites in exchange for the United States' pledge not to invade Cuba and agreement to remove its nuclear missiles from Turkey.

The point is that both the United States and the Soviet Union engaged in careless, hasty, and poorly understood nuclear testing and behavior at the peak of the Cold War. However, in the wake of the Starfish Prime test, the K Project, and the Cuban Missile Crisis, world leaders and public opinion pressured both nations to ratchet down their nuclear tests and threats. According to *Smithsonian*, "Both the Soviets and the United States conducted their last high-altitude nuclear explosions on November 1, 1962. It was also the same day the Soviets began dismantling their missiles in Cuba. Realizing that the two nations had come close to a nuclear war, and prompted by the results of Starfish Prime and continuing atomic tests by the Soviets, President John F. Kennedy and Premier Nikita Khrushchev signed the Partial Test Ban Treaty on July 25, 1963, banning atmospheric and exoatmospheric [outside the atmosphere] nuclear testing."[23]

For a brief time, at least, the treaty slowed the arms race between the Americans and the Soviets. Now, though, both the Soviet Union and the United States knew how to create a nuclear EMP weapon.

Starting in 2017, media interest in a nuclear E M P attack started to spike. This sharp increase was likely due to the West's tensions with North Korea, which as of this writing, has missiles capable of putting a nuclear weapon in low-Earth orbit. Because of the media reports regarding a nuclear E M P attack, more people are beginning to understand it.

However, most people might be surprised to learn that there are also nonnuclear E M P weapons, even though these weapons date to the early days of the Cold War. In 1951 Andrei Sakharov, a Soviet nuclear physicist and Nobel laureate, developed the concept for generating a nonnuclear E M P.[24] Sakharov was also the designer of Soviet thermonuclear weapons, but he later became an outspoken advocate of civil liberties and civil reforms in the Soviet Union and of stopping nuclear proliferation. Sakharov's public stance on these issues led to his being awarded the Nobel Peace Prize in 1975.

There are different types of nonnuclear E M P generators, and nations are keeping their latest designs top secret. In all cases, the E M P generated by a nonnuclear device is about a million times smaller than those created by a nuclear weapon; however, they can be instrumental in battle. Following are the three classes of nonnuclear E M P weapons, including examples of their use.

Flux Compression Generators: The most common nonnuclear E M P generators use flux compression (see appendix D for two articles that explain how they operate). These types of E M P generators have many configurations, some of which remain secret. Their most important aspect is they use a conventional explosive or compressed air to concentrate a magnetic field, eventually generating an intense E M P capable of rendering an adversary's electronics in a localized area useless. Ideally, that localized area may be the adversary's command-and-control headquarters or the electronic control systems critical to the operation of a weapon.

These E M P weapons vary in size. Some are large devices, carried by trucks or vans; others are smaller and carried by bombs, drones, and missiles.

Electromagnetic Pinch Devices: These devices use the "pinch effect" to create an EMP. (Appendix D lists an article that explains the pinch effect fully.) According to the editors of *Encyclopaedia Britannica*, "When an electric current is passed through a gaseous plasma, a magnetic field is set up that tends to force [pinch] the current-carrying particles together."[25] This concentrates the magnetic field. While this definition discusses using a plasma—that is, a hot ionized gas, or the fourth state of matter—to achieve a pinch effect, it is possible to use other conducting mediums, such as solid or liquid metal.

Nuclear physicists first used pinch devices to control nuclear fusion. In addition, the EMP created by a pinch device can be projected significant distances using a parabolic dish to focus on a distant target. If you watched the movie *Oceans 11*, you might recall the crew used a pinch device to knock out the power to Las Vegas. That device is probably Hollywood science fiction, or maybe it is not. We do not know because world militaries are silent on the capability of their pinch device weapons.

Marx Generators: Erwin Otto Marx, a German electrical engineer, first described this nonnuclear device in 1925.[26] These devices generate a high-voltage pulse from a low-voltage DC supply (i.e., a direct current, similar to a flashlight battery).

Sandia National Laboratories in Albuquerque uses a bank of thirty-six Marx generators to generate X-rays in its Z machine. According to its website, "Sandia's Z machine uses electricity to create radiation and high magnetic pressure, which are both applied to a variety of scientific purposes ranging from weapons research to the pursuit of fusion energy."[27] In another application, the high-voltage pulse from a series of Marx generators can serve as the ignition switch for a thermonuclear weapon.[28]

As the name implies, a nonnuclear EMP weapon, also known as an e-bomb, generates an EMP using conventional methods.[29] Therefore, nuclear treaties do not restrict their use. As the examples here demonstrate, alone and in combination with other weapons systems, nonnuclear EMP weapons can have devastating effects.

I consider nonnuclear E M P weapons as tactical weapons because they result in localized destruction. For example, a nonnuclear E M P can render a building housing the headquarters of an adversary, including its electronic and communications systems, dark. During warfare, the loss of situational awareness and communications is a recipe for defeat.

I consider a nuclear E M P as a strategic weapon, since it may result in mass destruction. Any armament that puts an entire nation in darkness is going to cause the death of many people.

An E M P weapon's destructive force impacts electrical and electronic systems but does not affect humans directly. For example, the people of Hawaii felt no physical ill effects from the Starfish Prime test, but their phones, radios, and televisions went dead. Nevertheless, you may have read articles that claim a 90 percent mortality rate in the first year following a strategic E M P attack, while other pieces are mostly dismissive of its effects.[30] Which are correct?

To address the question, we need to examine how a nuclear E M P weapon would be lethal to humans. According to *Washington Examiner*, which featured an article based on the U.S. Air Force's Air University 2018 report, an E M P attack that affected the entire United States would result in the following:

Ninety-nine nuclear reactors would probably melt down without electricity to control them.

About four million people located around the nation's nuclear plants would need to relocate due to released radiation.

Military and commercial jets might potentially crash.

It would cut off military bases, making military coordination difficult.

Civil unrest would start in "hours."

G P S could go dark.

Failures may include the long-term loss of electrical power, which would lead to problems with sewage, fresh water, banking, landlines, cellular service, and motorized vehicles.

It would require eighteen months or more to replace key elements of the electric grid damaged or knocked out by the EMP.[31]

The effects of a strategic EMP have the potential to break down society. People on life-support equipment would die. Without the capability to make 911 calls, no one could summon emergency medical technicians, firefighters, and police. Furnaces and air conditioners would cease working, allowing extreme temperatures to add to the EMP's lethality. No doubt, many people would perish. However, the previously cited statistic of a "90 percent mortality rate in the first year" arose during a theoretical discussion during a congressional hearing in 2008 and was based on a prepublication copy of the novel *One Second After* by William R. Forstchen.[32]

As far as my research reveals, no one has substantiated a number of projected fatalities of an EMP attack on the United States. I think a great deal would depend on how effective the attack is, as measured by the extent of the damage done to the electrical and electronic circuits in the country.

Speaking as a technologist, I am simplifying here and only providing a high-level overview in saying the effectiveness of a nuclear EMP attack will depend on four major factors. We must however consider these factors as guidelines, not absolutes, since aboveground nuclear testing ceased in 1962, and data regarding nuclear EMPs is limited. With these caveats, we should consider these four factors:

1. The size of the weapon, as measured in kilotons or megatons of TNT

2. The altitude of the detonation

3. The gamma ray output of the explosion

4. The detonation's interaction with the earth's magnetic field

Having touched on items 1 and 2 previously, let us discuss items 3 and 4. A nuclear blast produces gamma radiation, which is the radioactive emission from unstable atoms stripped of their electrons; it is the most energetic electromagnetic radiation.[33] When the explosion occurs at high altitude (i.e., about two hundred to three hundred miles above the earth's surface), the burst of gamma radiation ionizes atoms by stripping their electrons. This detonation occurs in the stratosphere, which is the second layer of the earth's atmosphere. As the free electrons interact with the earth's magnetic field, it produces a stronger EMP.[34] By contrast, if the detonation occurs near the earth's surface, the atmosphere blocks the effects of gamma radiation, and the EMP affects a smaller area around the blast.

Both the United States and Russia know these factors from firsthand experience, and other nuclear-weapon states such as China do from open source literature. They also understand that launching a nuclear EMP attack is equivalent to starting a nuclear war. As discussed in chapter 2, given that a nuclear war is not winnable, the United States and Russia have refrained from using their nuclear weapons. However, to my mind, the threat of a nuclear EMP attack remains. Let me explain.

North Korea currently has a primitive nuclear capability compared to that of the United States. It also has developed ballistic missiles that could hit the continental United States and its allies in the Pacific region. The U.S. Terminal High Altitude Area Defense system stationed in the region is capable of thwarting a state's launching a handful of nuclear-tipped ballistic missiles.[35] However, as the U.S. Congress noted in a 2017 report, North Korea could employ the "doomsday scenario" to launch an EMP attack on the United States in novel ways:

> North Korea could make an EMP attack against the United States by launching a short-range missile off a freighter or submarine or by lofting a warhead to 30 kilometers burst height by balloon. While such lower-altitude EMP attacks would not cover the whole U.S. mainland, as would an attack at higher-altitude (300 kilometers), even a balloon-lofted warhead detonated at 30 kilometers altitude could blackout the Eastern Electric Power Grid that supports most of the population and generates 75 percent of U.S. electricity.

Or an EMP attack might be made by a North Korean satellite, right now.

A Super-EMP weapon [a weapon that produces massive gamma radiation] could be relatively small and lightweight, and could fit inside North Korea's Kwangmyongsong-3 (KMS-3) and Kwangmyong-song-4 (KMS-4) satellites. These two satellites presently orbit over the United States, and over every other nation on Earth—demonstrating, or posing, a potential EMP threat against the entire world.[36]

This statement says to me that the threat of an EMP attack is real. Moreover, North Korea threatens to use its nuclear arsenal. According to a headline from a 2017 article in *Newsweek*, "North Korea Threatens Nuclear War with U.S., but 'Loves Peace More Than Anyone Else.'"[37]

Turning our attention to Iran, it has intermediate-range ballistic missiles that could be launched from a ship as part of a doomsday attack; however, it does not have a nuclear weapon, as of this writing. Nevertheless, with the withdrawal of the United States from the Joint Comprehensive Plan of Action (commonly known as the "Iran deal"), Iranian president "Hassan Rouhani said in a televised speech in 2019 that Iran would reduce its 'commitments' to the Joint Comprehensive Plan of Action, or JCPOA, but would not fully withdraw, amid heightened pressure from the U.S. in recent weeks."[38]

As a result, Iran is again enriching uranium above the levels set in the JCPOA.[39] What does this mean? Uranium found in nature is not suitable as nuclear fuel, and less than 1 percent of one isotope found in naturally occurring uranium, U-235, is adequate. Enrichment refers to concentrating the U-235 to higher levels. Concentrations over 90 percent constitute weapons-grade material. If Iran's production continues, eventually it will have enough material to begin assembling atom bombs. The technology for this primitive nuclear weapon is well known.

Many of my colleagues are dismissive of the EMP threat posed by North Korea and Iran, arguing the doctrine of mutually assured destruction is a sufficient deterrent. They also claim a state is more likely to target a city than to attempt to launch an EMP. I disagree. I think North Korea presents a clear and present danger and that Kim Jung-un, its leader, will initiate the doomsday scenario if he fears his own regime is doomed. Since he may not have enough missiles or weapons to inflict massive damage on U.S. cities, he may include an EMP attack as part of his strategy.

Iran cannot launch a doomsday scenario, as of mid-2019, but this situation is likely to change in the future. I am concerned Iran's ideological beliefs will supersede the doctrine of MAD, leading to a kamikaze mentality.

President Richard Nixon conducted foreign policy by attempting to convince enemy leaders he was irrational and volatile. Nixon's goal was to intimidate them. However, Nixon was acting. North Korea and Iran are not.

Abnormal stress and fears may lead to irrational attacks by North Korea and Iran. The doomsday option is on the table, and we need to deal with it.

Mitigating Factors

Although I have not discussed a solar flare–induced EMP in any detail, as previously mentioned, one hits the earth about every hundred years. The last significant one occurred in 1859 and set telegraphs sparking.[40] To a certain extent, you may say we are overdue for another.

Are we preparing for a potential EMP attack or even a solar flare–induced EMP? To a limited extent, we are. For example, President Trump's March 2019 Executive Order on Coordinating National Resilience to Electromagnetic Pulses states the threat and the role of the federal government in addressing it: "An electromagnetic pulse (EMP) has the potential to disrupt, degrade, and damage technology and critical infrastructure systems. Human-made or naturally occurring EMPs can affect large geographic areas, disrupting elements critical to the Nation's security and economic prosperity, and could adversely affect global commerce and stability. The Federal Government must foster sustainable, efficient, and cost-effective approaches to improving the Nation's resilience to the effects of EMPs."[41]

The executive order requires the secretary of homeland security to develop a plan by June 2019 to "mitigate the effects of EMPs on the vulnerable priority critical infrastructure systems, networks, and [national] assets." Further, the order states, "The Secretary of Homeland Security shall implement those elements of the plan that are consistent with Department of Homeland Security authorities and resources, and report to the APNSA [assistant to the president for national security affairs] regarding any additional authorities and resources needed to complete its implementation."[42]

The plan is not in the public domain, and I do not know the status of its implementation. The good news is that the federal government is beginning to address the EMP issue, and the executive order appears comprehensive. The bad news is that all work is in the early stages, and how long it will take to EMP harden the U.S. electrical grid and other critical elements of society is not clear.

In this overview of EMP weapons, the most significant point is that they represent a grave potential threat that the federal government is beginning to address. To date, no one has launched an EMP attack.

The next class of directed-energy weapons, which we address in chapter 7, "Cyberspace Weapons," does not represent a potential threat but actual current acts of war. The United States, Russia, China, and North Korea are routinely launching cyberattacks that can be quite destructive. I have firsthand experience. My company was the victim of a cyberattack that nearly drove us out of business. Unfortunately, the United States and its potential adversaries engage in cyberattacks every hour of every day. Also, the U.S. military is dealing with adversarial electronic warfare, which includes communications jamming, for example. Cyberspace has become the new battlefield, and everyone is at risk of becoming a victim.

7

Cyberspace Weapons

We live in a world where all wars will begin as cyber wars. . . . It's
the combination of hacking and massive, well-coordinated
disinformation campaigns.

—JARED COHEN

O n Tuesday, November 8, 2016, like many Americans, I went to
my designated polling place to cast my vote for president. If
you believed the forecasters, Hillary Clinton led in most polls,
suggesting that voting in the election was a formality. That night I
watched CNN's election coverage, as was my tradition. I expected
that early in the evening, CNN would be calling it for Clinton. Before
reporting returns, CNN quoted a spokesperson for the Trump campaign as saying he needed a "miracle" to win.

By 9:00 p.m. eastern time, returns did not look good for Clinton.
By 2:47 a.m., Trump got the miracle his campaign team sought. CNN's
Wolf Blitzer made this historic announcement: "Donald Trump wins
the presidency. The business tycoon and TV personality, capping
his improbable political journey with an astounding upset victory."[1]

Donald Trump won thirty states and the decisive Electoral College, 304 electoral votes to Clinton's 227. When the final count was
in, Hillary Clinton won the popular vote by more than 2.8 million
votes, meaning most Americans wanted Clinton to be their president. However, that is not how U.S. elections are determined. Was
the election rigged?

The answer depends on how you define "rigged." According to FiveThirtyEight, a website that focuses on opinion poll analysis, "Hillary Clinton would probably be president if FBI [Federal Bureau of Investigation] Director James Comey had not sent a letter to Congress on Oct. 28. The letter, which said the FBI had 'learned of the existence of emails that appear to be pertinent to the investigation' into the private email server that Clinton used as secretary of state, upended the news cycle and soon halved Clinton's lead in the polls, imperiling her position in the Electoral College."[2]

Many assert the Comey letter was the tipping point. His letter shifted the race three or four percentage points toward Trump, leading Clinton to lose Michigan, Pennsylvania, and Wisconsin by less than one point.[3]

However, it would be an oversimplification to state the Comey letter alone was responsible for Clinton's loss. Allegations that Russia interfered in the 2016 U.S. presidential election kicked off multiple investigations involving the Central Intelligence Agency (CIA), the FBI, and the National Security Agency. The January 2017 Office of the Director of National Intelligence report, using information gathered by those agencies, stated with "high confidence" that the Russian government conducted a sophisticated campaign to influence the recent election in favor of Donald Trump.[4] According to the Council on Foreign Relations, a U.S. nonprofit think tank specializing in U.S. foreign policy and international affairs,

> U.S. authorities say Russian agents hacked into computer systems associated with both major U.S. political parties. They are believed to have stolen thousands of emails from leading Democratic Party figures in early 2016 and leaked them online weeks ahead of the party's national convention in July. Russian military intelligence "used the Guccifer 2.0 persona, DCLeaks.com, and WikiLeaks to release U.S. victim data obtained in cyber operations publicly," U.S. intelligence agencies said in the January 2017 report. The leaked documents contained correspondence described by the *Washington Post* as "an embarrassing look at Democratic Party operations."[5]

This excerpt provides a valuable insight. Russia used cyber methods to attack the computer systems of both of the major U.S. political parties and used the information gleaned to discredit only the Democratic Party. Later during the election, the leaks would pointedly discredit Clinton and favor Trump. Russian perpetrated a disinformation campaign, using divisive and false information, to influence the outcome of the U.S. presidential election. According to Special Counsel Robert Mueller's investigation of this case and to the subsequent indictments of participants in 2018,

> Defendants [thirteen Russian intelligence operatives], posing as U.S. persons and creating false U.S. personas, operated social media pages and groups [including Facebook, Twitter, and YouTube] designed to attract U.S. audiences. These groups and pages, which addressed divisive U.S. political and social issues, falsely claimed to be controlled by U.S. activists when, in fact, they were controlled by Defendants. Defendants also used the stolen identities of real U.S. persons to post on ORGANIZATION-controlled social media accounts. Over time, these social media accounts became Defendants' means to reach significant numbers of Americans for purposes of interfering with the U.S. political system, including the presidential election of 2016.[6]

In keeping with this chapter's goal to understand cyberspace weapons, let us explore how the Russians hacked into the political campaigns of both parties. Get ready for a spy story, complete with Bitcoin payments, malware cyberattacks, and email spear phishing, the fraudulent practice of sending emails ostensibly from a known or trusted sender to obtain confidential information.

Let us start with the email spear phishing. Almost all of us with email accounts receive fraudulent emails that try to trick us into sharing passwords and personal information. As an author of books on artificial intelligence and high-tech weapons, I get several daily. In the case of the Democrats, Clinton's campaign chair, John Podesta, fell for a classic "reset your password" email that appeared to come from Google. The email popped into Podesta's inbox around March 19, 2016, but came from Aleksey Lukashev, a senior lieutenant in Russian military intelligence. Resetting the password occurred on a Russian-created phishing website made to look as if it were a Google page. The email also contained an embedded link, which covertly opened Podesta's account to a hacking team in Russia. The *Los Angeles Times* reports, "Two days later, the Russian cyber thieves stole—and later leaked—more than 50,000 of Podesta's private emails, incalculably undercutting Clinton's bid for the White House."[7]

Next, Russia hackers used malware, which is software that enables unauthorized access to a computer, on April 6, 2016, to send an email to thirty Clinton staffers. The email appeared to come from another staffer, whom the recipient viewed as a trusted source, and contained a fake Excel file that when opened, installed Russia's malware, a hacking tool called X-Agent.[8] A month later, the Russians stole data from thirteen computers of the Democratic Party in a single day and routed it through a leased server in Arizona, paid for with Bitcoin, a cryptocurrency.

Why use Bitcoin? One of the trickiest parts of intelligence operations is conducting financial transactions. Spies cannot use traditional electronic payment networks and checks since these instruments leave a trail of evidence. Before Bitcoin's emergence as a cryptocurrency in 2009, spies needed to haul around suitcases full of cash for significant operations.

To understand Bitcoin, we need to define "cryptocurrency." According to *Merriam-Webster*, a cryptocurrency is "any form of currency that only exists digitally, that usually has no central issuing or regulating authority but instead uses a decentralized system to record transactions and manage the issuance of new units, and that relies on cryptography to prevent counterfeiting and fraudulent transactions."[9]

Therefore, Bitcoin is independent of banks and nations. You can think of it as a way one person can enter into a transaction with another by paying in a currency they both accept. Here is how that works. A person first establishes a Bitcoin "wallet"—a program where the cryptocurrency is stored—and then uses a secret piece of data called a private key to access the wallet and sign transactions, providing a mathematical proof that they have come from the owner of the wallet.

Initially, the Russians bought some Bitcoins on peer-to-peer exchanges, where buyers and sellers can interact directly and incognito. Since the Russian ruble was weak at the time, which I estimate to be 2014 through 2016, I think they used gold to purchase the Bitcoins.[10] Next, according to the *New York Times*, "the Russians also created Bitcoins themselves through the process known as mining, the [Mueller] indictment said. With mining, computers compete to unlock new Bitcoins by solving difficult computational problems. This requires expensive equipment and lots of electricity, but that was apparently not a hindrance to the Russians."[11]

Spies love using cryptocurrency because it allows them to move millions of dollars across the world anonymously without the approval of any traditional financial institution. Paying in Bitcoin also enables all parties to hide their identities. The *New York Times* reports, "While the Russians accused of attacking Ms. Clinton's campaign also used traditional currencies, the [Mueller] indictment said they had 'principally used Bitcoin when purchasing servers, registering domains and otherwise making payments in furtherance of hacking activity.'"[12]

This operation perpetrated by Russia was extensive, involving at least the thirteen Russian operatives indicted by Mueller and an unknown number in Russia. According to the indictment, the latter group included the Internet Research Agency, located in St. Petersburg, that focused on spreading misinformation on social media. This group started its social media misinformation campaign in 2014 and reportedly was funded by Yevgeny Prigozhin, a businessman with close ties to Russian president Putin. Two of Prigozhin's other companies were also involved.[13]

In June 2017 U.S. Department of Homeland Security (DHS) officials stated that hackers linked to the Kremlin attempted to infiltrate election-related computers in more than twenty states but did not tamper with the vote count. The DHS assessed they were scanning the systems for weaknesses.[14] It is unclear if they found any.

The report from the Office of the Director of National Intelligence states, "We assess Russian President Vladimir Putin ordered an influence campaign in 2016 aimed at the U.S. presidential election. Russia's goals were to undermine public faith in the U.S. democratic process, denigrate Secretary Clinton, and harm her electability and potential presidency. We further assess Putin and the Russian Government developed a clear preference for President-elect Trump."[15]

To summarize Russia's 2016 presidential election meddling: President Vladimir Putin ordered Russian operatives in the United States and Russia to hack into the computers of both of the major U.S. political parties and use the information to discredit the Democrats and Hillary Clinton. Because of the Russians' "clear preference" for Trump, they withheld any potentially damaging information on the Republicans and Trump.

Why did the Russians favor Donald Trump? I was unable to find a definitive answer to that question.

Russia's Sophisticated Election Tampering

Special Agent Adrian Hawkins of the FBI called the Democratic National Committee (DNC) in September 2015. Although his information was urgent and vital concerning at least one hacked computer in the DNC network, the operator simply routed his call to the help desk. (If you or I had made a similar call, we would have landed in the same place.) Eventually, he spoke to Yared Tamene, a tech-support contractor at the DNC who thought Hawkins's call might be a hoax. Told that the perpetrators were the "Dukes," a cyberespionage team linked to the Russian government, Tamene made a cursory search of the DNC's computer system logs to look for such a cyber intrusion. Despite repeated follow-up calls by Special Agent Hawkins, Tamene did little to uncover evidence of the cyber intrusion.[16]

For the first time in U.S. history, the Russians used cyber espionage to compromise the 2016 presidential election, but unfortunately Tamene failed to detect it initially. In all fairness to Tamene, the FBI's approach by telephone communication did not help. According to the *New York Times*, "The low-key approach of the F.B.I. meant that Russian hackers could roam freely through the committee's network for nearly seven months before top D.N.C. officials were alerted to the attack and hired cyberexperts to protect their systems."[17]

The Russian cyberattack appeared to experts as highly sophisticated, to the point that *Politico*, an American political journalism company, ran this headline on July 5, 2017: "Democrats: Did Americans Help Russia Hack the Election?" However, there is no evidence of any Americans' involvement.[18]

What baffles security experts is the Russians' sophistication; they appeared to know which vulnerable Democratic candidates to exploit and what leaked information would court swing voters. According to the *New York Times*, "The fallout included the resignations of Representative Debbie Wasserman Schultz of Florida, the chairwoman of the D.N.C., and most of her top party aides. Leading Democrats were sidelined at the height of the campaign, silenced by revelations of embarrassing emails or consumed by the scramble to deal with the hacking. Though little-noticed by the public, confidential documents taken by the Russian hackers from the D.N.C.'s sister organization, the Democratic Congressional Campaign Committee, turned up in congressional races in a dozen states, tainting some of them with accusations of scandal."[19]

Russia's cyberattack was directed at the core of what makes us a nation—namely, our democracy. If any foreign power can manipulate our presidential elections, then we no longer have a democracy. With this in mind, let us define a cyber weapon.

What Is a Cyber Weapon?

Various reports and newspaper articles will use the term "cyber weapon" loosely. For that reason, I think it is necessary to provide a definition as it relates to this chapter: a cyber weapon is computer software or hardware used to disrupt the activities of a state or organization.

With this description in mind, let us discuss the first-ever cyber weapon, Stuxnet, arguably the most infamous of all.[20]

In January 2010, Iran's centrifuges, machines used to enrich uranium, at the Natanz nuclear facility began to burn out at an unprecedented rate. Both the International Atomic Energy Agency inspectors who were visiting Natanz at the time and the Iranian technicians responsible for the centrifuges were baffled.[21]

The computers used at the Natanz facility were isolated from the internet. To the Iranians, this provided a firewall, a part of a computer system or network designed to block unauthorized access against a computer virus attack. Then, in what appeared as an unrelated incident in June 2010, numerous Iranian computers mysteriously began crashing or malfunctioning. Iran attempted to fix the problem but failed. After repeated efforts, Iran decided to call in a security company. The Iranians contacted VirusBlokAda, a Belarusian security company, which assigned Sergey Ulasen to fix the problem.

Sergey Ulasen is a 2006 graduate from the Belarusian State Technical University with a degree in software development.[22] Ulasen, working from Belarus, provided phone support to a colleague in Iran. Together they discovered the Stuxnet virus. In an interview, Ulasen shared his experience:

> I must say it took a great deal of effort to find the cause of the anomalies. But we got there in the end. We eventually found the malware, and figured out its stealthy nature, strange payload and spreading techniques. . . .
>
> The complexity of Stuxnet's code and extremely sophisticated rootkit technologies [software tools that enable an unauthorized user to gain control of a computer system without being detected] led us to conclude that this malware was a fearsome beast with nothing else like it in the world.[23]

On June 17, 2010, Ulasen and his colleague documented the malware under another name.[24] Although they used a different name, this date marks Stuxnet's discovery.

Stuxnet is a malware computer program that replicates itself, similar to a virus, to spread to other computers. In the parlance of cyber weapons, it is a "computer worm."[25] It initially got into the computer systems in Iran's Natanz uranium-enrichment facility via a random worker's USB drive (a removable data storage device with an integrated universal serial bus plug).[26] Workers commonly use a USB drive to copy computer data to work on it at another location, such as on their home computer. (This is a high-tech version of taking home paper files in a briefcase.) The worker had no way of knowing the Stuxnet virus was lurking in the home computer and would embed itself on the USB drive. After working at home and storing the results on the USB drive, the worker inserted it into a computer at Natanz. Once inside the firewall, it spread from computer to computer.

Stuxnet is unique because unlike other computer viruses, it did not hijack computers or steal information. Instead, it sought to wreak havoc on the equipment the computers controlled. It jumped from the virtual world to become a physical reality.

Iran does not acknowledge the effects of the attack. However, some experts estimate that the Stuxnet worm destroyed close to a thousand uranium-enriching centrifuges, resulting in a 30 percent decrease in enrichment efficiency.[27] This cyber weapon set Iran's nuclear program back years.

At this point, you may be wondering who perpetrated the Stuxnet attack on Iran. Our typical archetype is a disenfranchised hacker, typically termed a "black hat," working secretly on a computer with ill intent. However, in the case of Stuxnet, nothing could be further from reality. The New York Times reports, "Though American and Israeli officials refuse to talk publicly about what goes on at Dimona [a complex of factories in Israel that produce nuclear arms], the operations there, as well as related efforts in the United States, are among the newest and strongest clues suggesting that the [Stuxnet] virus was designed as an American-Israeli project to sabotage the Iranian program."[28] If this report is true, the United States and Israel developed Stuxnet and jointly perpetrated one of the most devastating cyberattacks known to date.

Other cyber weapons are too numerous to address here. The Stuxnet discussion is one insightful example, and I want to highlight some crucial points.

Cyber is one battlespace in which the United States intends to maintain its leadership. To that end, in 2009 it established U.S. Cyber Command, and it was elevated to one of the unified commands of the Department of Defense in 2018.[29] Its mission is "to direct, synchronize, and coordinate cyberspace planning and operations to defend and advance national interests in collaboration with domestic and international partners."[30] In addition, the Defense Department's summary of its "2018 Cyber Strategy" coined a new phrase, "defend forward," which describes the department's efforts to "disrupt or halt malicious cyber activity at its source, including activity that falls below the level of armed conflict." The summary also states, "Our focus will be on the States that can pose strategic threats to U.S. prosperity and security, particularly China and Russia."[31]

NATO also recognizes cyberspace as an official war zone. A cyberattack on any NATO-protected state can trigger an Article 5 response—that is, a collective action by the twenty-eight-member organization against the perpetrator.[32]

Cyberspace warfare rages even as you read this book. Nations and non-state operators are continually waging cyberwar. Special Counsel Robert Mueller testified about it before Congress in July 2019. According to United Press International, a leading provider of news, "Cyber attacks have become more commonplace in recent years. The U.S. government has claimed multiple attacks over its computer networks from hackers in both Russia and China. The Islamic State terror group is also further developing its cyber-attack capabilities, intelligence officials say."[33]

The cyber battlefield surrounds you, and cyber weapons can traverse wirelessly at the speed of light. It is a terrifying new reality. However, in keeping with this book's focus on directed-energy weapons, we are moving on to discuss electronic warfare.

Electronic Warfare (EW)

Numerous articles on electronic warfare are available. A search using the key words "electronic warfare" on Google yields millions of results.[34] I think most people believe electronic warfare is a new military tactic perpetrated by the major powers; however, electronic warfare dates back over a century to the Russo-Japanese War of 1904–5. According to Army Technology, a news website on the defense industry, "In January 1904, the Eclipse-class protected cruiser HMS *Diana*, then stationed in the Suez Canal, is credited with making the first wireless interception in history, and so launching the era of signals intelligence."[35]

Among numerous examples of electronic warfare, here we explore two of them to increase our understanding. I am going to start with the one that saved a nation—namely, the Battle of Britain in the early years of World War II.

Electronic Warfare That Saved a Nation

After Nazi Germany conquered most of Europe, including France, it turned its attention to Britain. On July 10, 1940, the Nazi Luftwaffe began its bombardment of British army defenses on the southern coast of England. Winston Churchill said, "The Battle of France is over. The Battle of Britain is about to begin."[36]

Hitler calculated that the British would fold under the Luftwaffe's relentless attack, but British Hurricane and Spitfire fighter planes confronted each wave of Nazi Messerschmitt bombers. Far from folding, Britain's Royal Air Force (RAF) proved itself a formidable opponent. It became clear to Hitler that no invasion of Britain would be possible without first defeating the Royal Air Force. Initially, Germany attempted to destroy British runways and radar. However, the British made the Nazis pay for each attack by unfailingly fighting back.

Since the British fiercely resisted the Nazis' attack, Hitler decided to embark on a reign of terror, bombing London and other large cities. His goal was to break the British will to fight. He soon learned that this tactic would also prove futile.

On September 15, 1940, Germany initiated its most significant bombing attack on London with the hope it would clinch victory.[37] Again, the Luftwaffe faced a determined RAF, which scattered the German bombers, resulting in heavy Nazi losses. Historians mark this battle as the turning point in the Battle of Britain.

The Nazis believed that the British had them outnumbered in the skies above Britain. In vain, the Germans continued their bombing mission, but their raids began to slow. Hitler and his high command realized they could not defeat the Royal Air Force.

In reality, the British were outnumbered and frightfully low on both planes and pilots. How did they manage to win the Battle of Britain? They prevailed because of the bravery of their fighter pilots, the steadfast will of the British people and their leaders, and the use of electronic warfare.

The British did not have more planes than the Nazis, but they had a newly invented EW device, radar.[38] In 1935 the British Air Ministry asked Robert Alexander Watson-Watt, a physicist working at the National Physical Laboratory, to create a death ray using radio waves. If he had succeeded, it would have been what we now call a microwave weapon. Unfortunately, he failed but during his experiments discovered something that would prove equally valuable: he observed that transmitted radio waves would bounce back, as if they echoed, from any object encountered, including a plane more than two hundred miles away.[39] According to the History Learning Site, "Such a distance would give the RAF an early warning of an attack. As his [Watson-Watt's] work was done during the build-up to World War Two, such an invention was invaluable to the RAF that believed it was significantly weaker than the Luftwaffe. During the Battle of Britain, the Germans lost the element of surprise."[40] In 1940 this system was called radio detection and ranging, or radar.

Because the British used radar, the RAF was able to intercept the Germans' attacks and appear to be a more significant force than it really was. Losing the element of surprise proved devastating for the Nazis. For every British plane lost in combat, the Germans lost almost two.[41] Also, since radar guided them to the incoming Nazi bombers, the British pilots wasted no fuel hunting them and could stay in the air longer. Once airborne, they attempted to stop the Nazis before they could reach London.

Hitler's assault on Britain lasted about a year and ended in May 1941. Hitler stopped bombing the British because he needed his Messerschmitts to invade Russia.

While the Battle of Britain was one of the best examples of electronic warfare in history, electronic warfare has significantly evolved since then. Therefore, we should define the phrase before proceeding further.

What Is Electronic Warfare?

Electronic warfare involves using directed energy to control the electromagnetic spectrum—for example, radar and radio transmissions—to deceive or attack an enemy or to protect friendly systems from similar actions.[42]

Electronic warfare has subdivisions, which I have simplified to three:

1. Electronic attack, such as jamming an adversary's radio or satellite communications

2. Electronic protection, such as preventing an adversary's electronic attack

3. Electronic warfare support, such as using radar to detect potential threats[43]

A Recent Example of Electronic Warfare

You would think that since the United States spends about ten times more on military defense than Russia does, it would be superior in electronic warfare, but it is not. While the U.S. military dedicated decades to counterterrorism operations, Russia modernized its military with a particular focus on asymmetrical warfare, achieving warfare dominance in areas of U.S. weakness.

Experts agree that Russia has taken a massive, unexpected leap forward in its EW capabilities, which it uses to degrade the information environment of U.S. forces.[44] *Business Insider* reports, "After flexing its electronic warfare muscles during the annexation of Crimea following the 2014 Ukrainian revolution, the Russian military ramped up EW testing in war-torn Syria, disabling U.S. communications networks . . . in what then-U.S. Special Operations Command chief Gen. Raymond Thomas called 'the most aggressive EW environment on the planet from our adversaries.'"[45]

While Russia's EW capability may appear as a sudden surprise, it has its roots in the 2008 Russo-Georgian War. During that war, Russia's air force was unable to suppress Georgia's air defense systems through jamming efforts, costing Russia numerous aircraft losses in battle.[46] Learning from this experience, Russian president Putin ordered the modernization of at least 70 percent of Russian EW equipment by 2020.[47] Additionally, Russia formed dedicated EW units in 2009, with specialists proficient in EW tools. To my eye, it is not surprising that a decade later, Russia has a formidable EW capability that allows it to engage in noncontact operations that jam, blind, and disrupt an adversary without resorting to conventional conflict. The *Business Insider* reports, "In March 2019, the Norwegian government claimed it had proof that Russian forces actively disrupted GPS signals during Trident Juncture, the largest NATO war games since the end of the Cold War conducted around Northern Europe and the Arctic in late 2018."[48]

These types of engagements fit well with Russia's doctrine to blur the lines between peace and war. Syria's civil war also emerged as an ideal war zone for Russia to use and refine its EW capabilities.[49]

Why Electronic Warfare Is Critical

Whoever controls the electromagnetic spectrum will win the next major conventional conflict. I recognize this is a bold statement, even extraordinary. I also agree with Carl Sagan, "Extraordinary claims require extraordinary evidence."[50] Therefore, let me justify my assertion.

The term "electromagnetic spectrum" is used to describe all the electric and magnetic radiation that exists. Almost all of the electromagnetic spectrum, such as radio waves and X-rays, is invisible. However, it also includes visible light, which makes up a tiny fraction (0.0035 percent) of it.

I recognize that the electromagnetic spectrum is difficult to visualize. To assist in visualization, consider this example. In school, you may have studied the hydrogen atom, which has a nucleus of one positively charged proton surrounded by one negatively charged electron. When the electron is at rest, it generates an invisible electric field because it is a charged particle, and when it moves, it starts producing a magnetic field. These are fundamental properties of the physical world. In 1820 Hans Christian Ørsted observed that he could deflect a compass needle by switching a nearby battery's electric current on and off.[51] This experiment confirmed the connection between electricity and magnetism. Thus, we can include all wavelengths, from long radio waves to short gamma rays, in the electromagnetic spectrum.

Today's U.S. military makes extensive use of the electromagnetic spectrum. For example, it relies on GPS signals for navigation and targeting for aircraft carriers and cruise missiles, respectively. In addition to GPS, the U.S. military uses a wide range of other satellites. Some satellites are for surveillance. Timely information on the location of the enemy is critical and can act as a force multiplier, as radar did in the Battle of Britain. Other satellites are for communications, which serve as the foundation of military coordination.

One example that uses most aspects of the electromagnetic spectrum is the U.S. Navy's Aegis combat system, which uses integrated ship-to-ship communications, radar, and computers to guide missiles to enemy targets. All of these capabilities traverse the electromagnetic spectrum. Now imagine what would happen if an adversary jammed the navy's Aegis combat system. Such an attack would make even an aircraft carrier strike group vulnerable. In a sense, the navy would have to fight "unplugged," as if it had lost electrical power to its weapons. This assertion may sound farfetched, but Russia is trying to accomplish this capability and has even erroneously claimed that it succeeded in 2014. The United States says Russia's claim is simply propaganda.[52] However, the story does indicate that Russia is likely working on the way to monitor and jam naval communications.

The Aegis combat system example demonstrates how vital control of the electromagnetic spectrum is to the U.S. Navy, but the same is true for other weapons systems of the U.S. military and of other world militaries. Adversaries continually engage in electronic warfare to listen to and jam each other's communications and satellite links.

The Awakening

The United States is aware that it needs to regain dominance in electronic warfare. The Pentagon plans to increase its focus to counter jamming and hacking, as well as EW protection for friendly forces. To accomplish these goals, the Defense Department initiated two strategies:

1. It set up a new task force, led by U.S. Air Force general and vice chairman of the Joint Chiefs of Staff Paul Selva, to produce an "updated electronic warfare strategy and roadmap" for Congress.

2. It primed defense contractors to prepare for the rapid prototyping of urgently needed electronic warfare weapons.[53]

Playing catchup can be costly. According to the Congressional Research Service, a public policy research arm of Congress,

Congress, in the FY2019 National Defense Authorization Act, and the Department of Defense (DOD) has identified electronic warfare (EW) as a critical capability supporting military operations to fulfil [sic] the current National Defense Strategy. . . .

[The] DOD requested at least $10.1 billion in FY2019 [fiscal year 2019], $10.2 billion in FY2020, and $9.7 billion in FY2021 for EW.[54]

These statements and budget requests say one thing to me: the U.S. military views the EW threat seriously and intends to prepare for it.

In the last few years, China has licensed the rights to publish two of my books—*The Artificial Intelligence Revolution* and *Nanoweapons*. Their upfront royalty advances were significant, and they conducted business ethically. China also offered me a hefty fee to speak on artificial intelligence at a conference held in China, but I politely refused the speaking engagement. I found the idea of going to China a scary proposition. I held a secret clearance for over twenty years while working on some of the United States' most advanced weapons. My concern was that the speaking engagement was a ruse. Let me explain.

Wired published an article in 2018 titled "China's 5 Steps for Recruiting Spies," and to explain my concern, I am going to use the steps outlined in the article. Step 1 is spotting an individual who can provide classified information. Step 2 is assessing what might motivate that individual to divulge the material. Typically these incentives come down to "money, ideology, coercion, and ego." Step 3 is developing a relationship. According to *Wired*, "Intelligence officers generally don't lead off by asking potential sources to betray their country or their employer." So in this third stage of espionage recruitment, "recruiters begin to ask for trivial requests or favors to establish rapport. As former CIA director John Brennan said last year, 'Frequently, people who go along a treasonous path do not know they are on a treasonous path until it is too late.'"[55] I believe that had I gone to China to participate in the conference, I would have found myself facing the tactics of step 3.

Step 4 is recruiting, which can be just a subtle as step 3. For example, it could be as simple as a conference chairperson requesting that I write a white paper (i.e., an authoritative report) on some critical aspect of artificial intelligence. A fee for such a document is routine. During my career, I wrote countless white papers—technology reports and proposed solutions for technological problems—and charged a fee for them, as was typical. I think if China had approached me to write a white paper, the proposed compensation would have been significant.

Finally, step 5 is handling the recruited spy. This step involves "encrypted communication tools, surreptitious cell phones, and emails left in draft folders."[56]

I think, after studying this subject, some spies do not realize they are spying. China's method for recruiting spies has nothing to do with what you see in James Bond movies, with their guns, physical dead drops (i.e., leaving information in a secret location), and in-person brush passes (i.e., passing along information as you brush next to someone).

Instead, China uses nontraditional spycraft tactics as well as cyber-attacks. A 2018 U.S. Department of Justice report delineates China's nontraditional spy recruitment methodology: "In October, the Department['s National Security Division and U.S. Attorneys' Offices] announced the unprecedented extradition of a Chinese intelligence officer, Yanjun Xu, who allegedly sought technical information about jet aircraft engines from leading aviation companies in the United States and elsewhere. To get this information, he is accused of concealing the true nature of his employment and recruiting the companies' aviation experts to travel to China under the guise of participating in university lectures and a nongovernmental 'exchange' of ideas with academics."[57]

That same Justice Department's report documents China's cyber espionage: "[China] managed a team of hackers to conduct computer intrusions against at least a dozen companies, a number of whom had information related to a turbofan engine used in commercial jetliners."[58] In addition, the report states that China recruits Chinese nationals as spies, targeting those individuals who work "as engineers and scientists in the United States (some for defense contractors)."[59]

You may wonder how effective this combination of nontraditional spycraft and cyber espionage is. The Justice Department's report states, "From 2011–2018, more than 90 percent of the Department's cases alleging economic espionage by or to benefit a state involve China, and more than two-thirds of the Department's theft of trade secrets cases have had a nexus [connection] to China."[60]

Newspaper stories and magazine articles about China's cyber threat appear on a routine basis. They seem to have a common theme: China is using cyber espionage to gain global dominance, and its government is responsible.[61] However, given the nature of cyber-attacks, proving its responsibility is impossible; it is only possible to trace those attacks as originating from China. Nonetheless, two things are apparent:

1. China sees cyber espionage as a way to advance beyond the United States economically and militarily.[62] Specifically, China's cyber espionage targets the U.S. diplomatic, economic, academic, and defense industrial base.[63]

2. The Chinese government—more specifically, Unit 61398 of China's People's Liberation Army (PLA)—is responsible for almost all cyberattacks originating in China.[64] According to the *New York Times*, "[A] 12-story building on the outskirts of Shanghai is the headquarters of Unit 61398 of the People's Liberation Army."[65]

China denies it engages in cyber warfare and accuses the United States of waging cyberwar against China, and the United States denies the charges. However, according to a 2019 article in the *New York Times*, "Chinese intelligence agents acquired National Security Agency hacking tools and repurposed them in 2016 to attack American allies and private companies in Europe and Asia."[66]

This excerpt highlights that both countries engage in cyber warfare. The *New York Times* characterizes it as a "digital wild West with few rules or certainties." It also demonstrates how difficult it is for even cyber-savvy nations, such as the United States, to guard its cyber tools.

In turning our attention to China's EW capabilities, in 2015 the PLA created the Strategic Support Force to centralize China's cyber and electronic warfare under a single organization. The Chinese view information dominance, the combination of cyber and electronic warfare, as crucial to winning conflicts.[67]

My reading of the 2019 Department of Defense's annual report on China's military and security developments confirms China's 2015 efforts. Furthermore, the DOD report adds, not only is China making significant investments in EW hardware and training but also its "EW strategy emphasizes suppressing, degrading, disrupting, or deceiving enemy electronic equipment" and GPS, surveillance, and communications satellites.[68]

This discussion completes my overview of China's cyber and electronic warfare thrusts. Let us now examine the future of cyberspace warfare.

The Future of Cyberspace Weapons in the Coming Decade

Cyber Weapons

Over the last decade, I have ordered my prescription refills by telephone. Initially, the computer-generated speech was robotic. In the last few years, the computer voice and interaction come across as almost human. What changed?

The machine now uses AI to interact, removing almost all traces of robotics. I think a "smart agent" (i.e., a computer algorithm that replicates a human specialist in a narrow field) is responsible. AI designers are making extensive use of smart agents. One example is any game on your computer or smartphone that provide options regarding the level of expertise at which you can play it. My phone's chess game has ten levels. I can usually win all games at level 1, most games at level 2, some games at level 3, and none at level 4. I suspect level 10 would be challenging to even highly ranked chess players.

The same development is happening for the computer's interface with humans when refilling a prescription. I suspect that newcomers to the system might think they are talking with a human operator, not a machine.

What is occurring in our everyday lives is also happening in cyber warfare. Most of today's cyberattacks depend on a programmer writing malware code and finding a way to insert it into a target's computer. With this approach, the malware has cockroach-level intelligence, similar to Stuxnet's: it knows how to hide, disguise itself, what to attack, and when. In the next decade, malware will incorporate AI. It will learn to impersonate people you trust, find ways through sophisticated firewalls, and elude detection. Most important, it will be persistent. It will never sleep or stop its attack. Such malware will be an autonomous weapon.

The U.S. military is aware of the potential threat that cyber weapons may become autonomous. To counter it, the Department of Defense Directive 3000.09 concerning "Autonomy in Weapon Systems" enables operators "to exercise appropriate levels of human judgment over the use of force" but does not apply to "autonomous or semi-autonomous cyberspace systems for cyberspace operations."[69] This exclusion means U.S. computer defenses against cyberattacks can act instantaneously without human supervision, which is necessary because cyberattacks outpace human reaction time.

Thus, the bottom line is the next leap in cyber warfare will be in making cyber weapons more autonomous, or artificially intelligent.

Electronic Warfare

The United States faces many challenges related to its dependence on space assets—such as GPS, surveillance, and communication satellites—and its use of the electronic spectrum, such as radar and radio communications. Russia leads in EW, and the United States is playing catchup. However, the United States is acutely aware of the EW challenges it faces and is allocating significant resources to regain EW dominance. Given the Department of Defense's budget allocations for 2019–21, I think the United States will regain the lead by 2025. In the interim, U.S. military planners are likely developing workarounds to fight unplugged if necessary.

I expect Russia to continue to push its lead in electronic warfare since it provides a way to best the United States in one area of conflict. However, the Russians are likely to confine their EW to proxy wars, such as in the Syrian civil war, and not confront the United States directly. The U.S. military is still a superior force with a full menu of conventional weapons to unleash on an adversary.

China, meanwhile, is likely to continue to consolidate its power in the Asia-Pacific region and do everything it can to make itself a formidable EW adversary while also playing catchup. Likewise, I doubt China wants to have an open conventional conflict with the United States, since the United States remains its largest trading partner as of this writing. My take is that China will prioritize its economy over its military aspirations in the coming decade.

The bottom line is Russia's lead in EW will erode through the mid-2020s as the United States focuses significant resources on regaining its supremacy in this critical technology. China will do all within its power to gain a stronger EW capability while continuing to court U.S. trade over military conflict.

Today, the United States and other technologically advanced countries are highly dependent on computers. In the future, that dependence will become total. Everything that enables you to survive will depend on a computer, from the water you drink to the food you eat. Computers will become as ubiquitous as electronics are today. Your car, work, finances, leisure, health care, and almost everything else will require them to operate. The internet of things (i.e., connecting everything to the internet) will engulf most of the world.

In that future time frame, imagine any nation that is unplugged—meaning all its critical computers malfunction or it loses electrical power—in the wake of a cyberspace or directed-energy weapons attack. That nation would fall apart, and its people would perish. Such an attack would have the same lethality as nuclear weapons. If this possibility concerns you, you are not alone. U.S. military planners share your concern. Fiction such as *Star Trek* inspired these weapons, and it is inspiring defenses against them. Part 3, "Shields Up, Mr. Sulu," may provide a small measure of comfort.

THREE

"SHIELDS UP, MR. SULU"

8

Directed-Energy Countermeasures

> We need to have a strong defense focused on areas
>
> that are in the greatest vulnerability.
>
> —BOB GRAHAM

I am taking martial arts classes from a seventh-degree black belt in jujitsu, which is Japanese for "gentle art." One of the most important things my instructor has taught me is to counterstrike an attack as early as possible. Therefore, when an attack is in its initial phase, you begin defending yourself. Most people, untrained in martial arts, wait until the attack is in full progress. Even then, many do not believe it is happening and, as a result, become a statistic.

This same concept forms a sound military defense strategy. For example, when a potential adversary cannot match U.S. offensive capabilities, it focuses on defensive strategies. This military strategy dates back to Sun Tzu, who was recognized for his work *The Art of War*. He wrote, "Appear weak when you are strong, and strong when you are weak. If your enemy is secure at all points, be prepared for him. If he is in superior strength, evade him. If your opponent is temperamental, seek to irritate him."[1]

In this chapter, we explore countermeasures that nations are taking to shield their weapons, such as missiles and planes, from directed-energy weapons. This strategy allows them to avoid a costly directed-energy arms race or, in the case of the United States, to buy time to gain dominance.

In what follows, be prepared for a classic game of cat and mouse. For every offensive directed-energy weapon that appears, nations, including the United States, are attempting to develop a countermeasure. To my mind, their strategy follows this simple premise: for every offense, there is a defense.

Laser Countermeasures

Lasers project highly focused energetic light at a target. When the beam reaches its target, three outcomes are possible: deflection, reflection, or absorption. The laser weapon is most effective when the beam is absorbed. If the beam is reflected or deflected, it means the target has thrown off some or potentially all of the laser's energy. Complete reflection and deflection would render the laser weapon useless and represent an ideal countermeasure.

In addition to deflection and reflection, targets can take passive countermeasures. For example, a missile could rapidly rotate, which spreads the heat, or achieve a higher acceleration, which quickly changes the targeting point.[2]

To understand reflection as a countermeasure, let us look at the approach China is taking.

China's Laser Reflective Coatings

China is investing in the research of coating materials, such as low-cost metals, rare earths, carbon fibers, and silver, whose physical and chemical properties reflect specific types of U.S. military laser beams. The phrase "specific types" represents a potential flaw in this defense. A different kind of laser or a laser that can change its beam to match the coating's absorption will defeat this countermeasure. Nonetheless, according to a 2014 article in the *South China Morning Post*, "Laser weapons like those developed by the United States pose little threat to the PLA—smog or no smog—because mainland researchers have pioneered coatings that can deflect beams and render them harmless, mainland scientists say."[3]

To my eye, this report is mostly propaganda. The problem is coatings require too many variables to align. China must finely tune coatings to reflect a specific laser beam's wavelength. Even the Chinese are aware of this pitfall. The *South China Morning Post* article includes a cautionary observation from Professor Huang Chenguang, a specialist in high-energy laser beams: "There were still a number of uncertainties regarding the effectiveness of the Chinese-developed anti-laser technology."[4]

The *Post* article was concurrent with the U.S. deployment of its first thirty-five-kilowatt laser weapon on the USS *Ponce*. This new weapon ushered in the era of directed-energy weapons, and I think the Chinese felt they had to diminish its importance.

The U.S. Navy's next laser, which Lockheed Martin is building, is likely to be about three to five times as powerful and may nullify the effects of reflective coatings.[5] It is reasonable to ask how it would do so. No missile coating is likely to be 100 percent efficient. Some absorption will occur. The intake of a small fraction of a high-energy laser means it will turn that minute surface area to molten metal plasma, or a hot ionized gas. The escaping plasma is likely to disrupt the coating in the surrounding area. Since U.S. military lasers are pulsed, shooting bursts every fraction of a second, the surface integrity at the target point may be lost quickly, allowing the next pulsed shot to destroy the target.

A simple way to visualize this process is to think about how rust spreads on a car. If hail, for example, causes a spot of the car's paint to chip off, rust will occur at that spot. In time, the corrosion will continue to spread underneath the paint, which will appear to slightly bubble before it eventually falls off.

China's coating countermeasure strategy has another flaw. In the early 2020s, the United States is likely to deploy microwave weapons along with the lasers. There is no guarantee a coating that reflects laser light will be effective against high-energy microwaves. Even so, in the face of all uncertainties, China is pursuing this approach as a low-cost countermeasure.[6]

Why is China pursuing a potentially flawed countermeasure? China may know something we do not know. For example, perhaps China knows how to make a coating that is totally efficient against a laser. While that may be one possibility, I doubt it. Having worked on many U.S. military weapon systems and satellites, I understand that the U.S. military explores all options and would not field a weapon if it was easily defeated. Given the U.S. Navy's and Army's contracts for high-power laser deliveries and deployment in the early 2020s, the U.S. military is apparently confident the lasers will perform against laser-coated countermeasures.

I also think the U.S. military is seeking higher-energy lasers to render passive countermeasures useless. In the case of a laser, absorbed energy destroys targets. If, for example, it is necessary to keep a 30-kilowatt laser on target for five seconds, then a 150-kilowatt laser will only require one second. Since a 150-kilowatt laser is the next laser weapon that the U.S. military is likely to deploy, one pulse may be all it takes to destroy a target.

Countermeasures for High-Power Microwave and Electromagnetic Pulse Weapons

High-power microwave (HPM) and electromagnetic pulse weapons destroy electronics using an electromagnetic field. With this commonality, I think discussing them together is appropriate.

As noted previously, the high-altitude detonation of a nuclear weapon generates a high-energy EMP—also known as a high-altitude EMP (HEMP)—that can affect areas as large as the continental United States. High-power microwave weapons generate far less energy, but they still can cause damage to unshielded electronics over a smaller area such as in a command center.[7]

In the case of a HEMP, a nuclear weapon produces the electromagnetic pulse. This type of attack requires a missile delivery system and nuclear weapons technology. An HPM weapon involves a dramatically lower level of expertise, using easily obtained electrical materials and chemical explosives costing less than $2,000, and will fit in a vehicle or even a suitcase. This type of HPM weapon is within the capability of most nations and some terrorist groups.[8]

Defending against HEMP and HPM weapons is problematic since power surges could cause peak currents of tens of millions of amps that can pass through a protective Faraday cage. Although elementary physics classes teach that a Faraday cage will block an electromagnetic field, that is only true in theory. In practice, a Faraday cage is not 100 percent efficient, and some of the electromagnetic field caused by a HEMP and HPM can penetrate it. The amount of penetration may be sufficient to cause electronic devices to burn out.[9]

At this point, you are probably wondering, What countermeasures are possible? According to a 2008 Congressional Research Service report, "Depending on the power level involved, points of entry into the shielded cages can sometimes be protected from electromagnetic pulse by using specially designed surge protectors, special wire termination procedures, screened isolated transformers, spark gaps, or other types of specially-designed electrical filters."[10]

I have hands-on experience in this area. A significant portion of my career was devoted to developing radiation-hardened integrated circuits, which, as discussed previously, can withstand high-radiation environments without shielding. Their exact specifications are classified. The U.S. military is acutely aware of the electronic damage radiation can cause, and the DOD invests significant funds in addressing it. As a result, crucial U.S. military equipment can operate either in space, where it experiences solar radiation (i.e., radiant energy emitted by the sun), or during an EMP attack.

Any detonation of a HEMP over the United States by an adversary would constitute a nuclear attack and invite a nuclear response by the United States and its NATO allies. However, a terrorist group with a suitcase-sized electromagnetic pulse device is more challenging to address. Although the area affected would be significantly less, a properly detonated EMP could result in a significant infrastructure loss, perhaps even of one of our electrical grids, and assessing the responsibility for such an attack may be difficult.

On a larger scale, nonnuclear pulse weapons can constitute the warhead of a missile or bomb. Reportedly, the U.S. Navy secretly used nonnuclear EMP warheads in the First Gulf War to disrupt and destroy Iraq's electronic systems. Conceivably, a rogue state or an organized terrorist group, using a truck or a car, could deliver a similar warhead against vulnerable infrastructure of the United States.[11] I think this scenario, as well as the possibility of a rogue state's HEMP attack, motivated President Trump to issue the Executive Order on Coordinating National Resilience to Electromagnetic Pulses in 2019. This effort to assess the risks of HEMP and HPM attacks, as well as of solar EMPs, on critical U.S. infrastructure would be the first step in defending against such events.[12]

Cyberspace Weapon Countermeasures

Cyberspace weapons are either cyber weapons, such as computer viruses, or electronic warfare weapons, such as those that jam GPS signals, radio communications, and radar surveillance, that seek to control the electromagnetic spectrum. To defeat them, we need cyber weapon countermeasures.

Cyber Weapon Countermeasures

Microsoft and other technology companies pay hackers to find security holes in their products. Microsoft began offering $100,000 "bug bounties" in 2013.[13] It announced at "the Black Hat USA 2019 hacking and security event in Las Vegas, that it has paid $4.4 million (£3.6 million) to hackers over the past 12 months."[14] Being a bug bounty hacker for Microsoft thus can be extremely lucrative. According to *Forbes*, "Hackers who can 'demonstrate a functional exploit' [in certain Microsoft products] ... can expect a $300,000 (£247,000) payday."[15]

This process may sound as if the company is hiring a thief to catch a thief, but it is essential to understand not all hackers are criminals. Some highly talented software coders are capable of hacking. Some earn bug bounties or work for U.S. government agencies, such the CIA. The intelligence agency's cyber officers develop cyber tools to protect U.S. data, gather intelligence, and launch cyberattacks. These software coders are engaged in legitimate work, although some of their work may be for U.S. government covert operations. If you ever watched the CBS drama series *Criminal Minds*, the character of FBI technical analyst Penelope Garcia, played by Kirsten Vangsness, is an example of a cyber coder engaged in legitimate work.

Cybercriminals are also software coders who are capable of hacking, but their intent is malicious as they engage in such activities as spreading viruses, gambling illegally, stealing money, or hijacking a victim's system to extort a ransom. For example, in May 2019 the *New York Times* reported, "For nearly three weeks, Baltimore has struggled with a cyberattack by digital extortionists that has frozen thousands of computers, shut down email and disrupted real estate sales, water bills, health alerts and many other services."[16]

Without going into more detail, large technology companies such as Microsoft offer the bug bounties to uncover security weaknesses in their systems so they can avoid what happened to Baltimore. Offering bug bounties is a crucial way to build cyber weapon countermeasures.

Government agencies also offer bug bounties to secure their systems. Doing so is the law in some cases. For example, in 2018 the U.S. Senate unanimously passed the Hack the Department of Homeland Security Act of 2018.[17] The DHS Act uses ethical hackers to identify vulnerabilities in the DHS networks. Other government agencies do likewise.[18] Using "white hat" (i.e., ethical and vetted) hackers is a crucial way the U.S. government secures its systems against cyberattacks.

The U.S. Defense Digital Service usually leads these operations and knows the bounty hunters do not remove the threat; they only uncover it. When a security loophole is revealed, fixing it can take up to $30,000.[19] However, money is not the main issue. The service is struggling to find capable coders to mitigate the software vulnerabilities. Even the contractor who built the system may be unable to develop a fix.

Now, let us turn our attention to electronic warfare countermeasures.

Electronic Warfare Countermeasures (EWC)

The United States is in a catchup mode relative to Russia's electronic warfare capabilities. As mentioned in chapter 7, the Department of Defense is budgeting more than $10 billion annually for the fiscal years 2019 and 2020 to regain its dominance in EW. That dominance will include having electronic warfare countermeasures against Russia's EW capabilities.

With the fall of the Soviet Union in 1991, the U.S. military largely abandoned EW. Following al-Qaeda's attacks on the United States in 2001, the U.S. military turned its attention to counterinsurgency, first in Afghanistan and later in Iraq. Over the next two decades, the U.S. military became the best counterinsurgency force in the world. Meanwhile, with Russia's invasion of Ukraine on August 22, 2014, enabled significantly by aggressive jamming of Ukrainian military communications, electronic warfare capabilities emerged as an urgent necessity for the U.S. military.

Fortunately, the U.S. military can leverage its military-industrial complex to achieve its ultimate goal—namely, EW dominance. With the massive allocations in the DOD budget, military contractors such as Lockheed Martin, Northrop Grumman, and Raytheon are vying to get lucrative contracts for next-generation EW devices.[20]

From experience, I can tell you that military contractors stay close to their counterparts in the DOD. Because of these relationships, the contractors are knowledgeable about DOD needs and can assess how well their capabilities fill those needs. As a result, a contractor may provide an unsolicited white paper, or an authoritative report giving information or proposals on an issue, that defines an unarticulated need and a contractor's resources to address it. However, if more than one contractor can address the need, the DOD will solicit bids via a request for proposal (RFP). When this happens, the contractor hopes that its white paper shapes the RFP to favor its strengths.

A proposal to an RFP must be responsive, meaning it has to meet all deliverables, cost, and schedule requirements. If the RFP specifies advanced technology that requires substantial development, the DOD may issue the RFP as "cost-plus," meaning the contractor can charge development costs plus a fixed profit of typically 8 percent.

In my experience providing radiation-hardened integrated circuit electronics, Honeywell was one of only two companies with the technology to build those products. Rarely would an RFP be listed that we did not know about beforehand. Therefore, in bidding on

DOD procurements, we competed against one company. Typically, relatively few contractors are qualified for DOD RFPs when they require advanced technology.

The DOD awards contracts based on the contractor's proposal that most closely meets the selection criteria in the RFP. If two contractors are equally capable and meet the selection criteria, the DOD awards the contract to the lowest bidder. At times, I had to decide Honeywell's final bid in response to a DOD RFP. That bid was so sensitive that only select employees, with a need to know, had access to it. Those proposals were "nail biters" and could substantially affect the business unit's profits and losses.

My point in sharing this is to demystify the DOD procurement process and delineate the criticality of having a robust military-industrial complex. Working together, defense contractors and their DOD counterparts enable the U.S. military to develop and deploy advanced weapons. This relationship is especially critical in developing electronic warfare countermeasures. To illustrate, let us examine the EW countermeasures of the U.S. Navy, Army, and Air Force.

U.S. NAVY EWC

To illustrate how vital the relationship is between the DOD and its military contractors, I am starting with the U.S. Navy because its long history in electronic warfare dates to the 1980s and the deployment of a missile defense system built by Raytheon and Hughes Aircraft that scanned the electronic spectrum for incoming missiles. Over the decades, the navy has upgraded its EW capability to go beyond missile detection to include ships, radio traffic, and other electronic signals.

With Russia's rapid advancement in EW, the navy awarded Lockheed Martin a $184 million contract for the ongoing production of Block 2 systems, which are part of the Surface Electronic Warfare Improvement Program. The navy's goal is to harden its warships against even Russia's hostile electronic warfare environment. According to Lockheed Martin, the upgrade will "expand upon the receiver/antenna group necessary to keep capabilities current with the pace of the threat and to yield improved system integration."[21] The upgrade also includes an open interface (i.e., a standard for connecting hardware to hardware and software to software) for the navy's critical combat systems. Once in place, the service believes it will be ready for twenty-first-century electronic warfare.

However, to gain dominance, the navy is taking EW one step further by soliciting proposals to take electronic warfare equipment that "goes to eleven" (meaning beyond a power dial's range of one to ten). Quartz, a business news organization, reports, "Reflecting the Navy's increasing use of small, unmanned ships and drones to augment and extend the capabilities of its existing force, the service branch is looking for ways to use a multitude of these devices to at once create an antenna capable of sending out radio frequencies with the power equivalent to those emitted from 'black hole jets' or 'gamma ray bursts' [celestial bodies that generate extreme power at incredibly high frequencies]."[22]

To jam a radio transmission, an adversary must be able to overpower it; thus, the navy seeks to increase the power and frequency of its electronic transmissions to a level that makes it impossible for an adversary to jam. I think of it as a swarm tactic applied to electronic warfare.

The navy is also exploring using autonomously controlled EA-18G Growler electronic-warfare aircraft to extend its EW reach while keeping manned aircraft out of harm's way.[23] If successful, this EW drone would provide an unmatched battlefield capability. Aircraft are extremely effective EW platforms due to their ability to reach high altitudes, which allows them to attack distant targets.

U.S. ARMY EWC

Initially ill-equipped, the U.S. Army found itself using large numbers of short-range jammers that were developed to disable radio-detonated roadside bombs as its primary EW tool.[24] However, money does solve some problems. With breakneck speed, engineers modified the army's short-range jammers into long-range jammers called Saber Fury. The army mounted Saber Fury on Strykers (eight-wheeled armored fighting vehicles) fielded by the Germany-based Second Cavalry Regiment and later mounted the system on Humvees as well.[25] It was a baby step in the right direction but still a long way from gaining electronic warfare dominance.

Next, the army produced a higher-power version of Saber Fury and installed it on a handful of heavier mine-resistant, ambush protected all-terrain trucks designated as electronic warfare tactical vehicles. But the army knew it had pushed Saber Fury about as far as possible. It was time to end the Band-Aid approach and develop new EW technology. The U.S. military's first attempt at integrating electronic and cyber warfare resulted in the Tactical Electronic Warfare System (TEWS), which was mounted on the larger and better-armored 8x8 (eight-wheel drive) Strykers. TEWS offered a better sensing and electronic attack capability, including a rudimentary ability to launch wireless cyberattacks.

The next improvement added cutting-edge signals intelligence technology to the system. The result gave the army in integrated EW and signals intelligence in one system, allowing the army to examine an adversary's communications and to conduct electronic warfare. As of this writing, the army has three of these systems and is using it in military exercises.[26]

TEWS represents a substantial beginning for the U.S. Army in taking significant steps toward its goal of combining EW and cyber warfare with signals intelligence by 2023. This system will result in a new form of digital war called multi-domain operations.[27] However, initial tests of this system revealed several challenges that the army must address, including operation on battery power limited to twenty minutes and unreliable communication from TEWS to the brigade.[28] If the army successfully overcomes these challenges, it will have the capability to jam and hack an adversary's electronics without detection.

U.S. AIR FORCE EWC

I saved the EW countermeasures of the U.S. Air Force for last because aircraft are the top-tier weapons in this form of combat. They owe this crucial position to one fact: a plane's altitude allows it to use EWC over a larger area than a ship or land-based vehicle could.

Although most military air forces use EWC, the U.S. Air Force can couple it with stealth, making it more effective in battle. The first job of aircraft EWC is to prevent radar tracking. Stealth aircraft are typically able to evade radar tracking, but when necessary, they use jamming and spoofing, or deceiving an enemy force's radar into displaying inaccurate information. Older U.S. aircraft, which are less stealthy, rely on EWC to jam or spoof enemy radar. The second and equally critical job of EWC is to defeat surface-to-air missiles or air-to-air missiles.

U.S. Air Force fighter planes and dedicated EWC planes carry jamming pods (i.e., a suite of electronics and antenna), with the latter having the most powerful systems. Dedicated EWC planes ensure the survivability and effectiveness of other aircraft and friendly ground forces. This capability enables fighters and bombers to penetrate enemy airspace more safely.

However, as with the other service branches, the air force's year-long 2018 study makes clear that it fell behind in this type of combat relative to its sophisticated adversaries.[29] Addressing this area of vulnerability is critical since air dominance is a strategic component of how the U.S. military fights wars. U.S. aircraft, more than any other fighting machine, rely on the electromagnetic spectrum for GPS, data links, communications, radar, and weapon capabilities. If the air force is unable to dominate the electromagnetic spectrum, its planes become as ineffective as World War II aircraft and could be easily defeated by a sophisticated adversary.

The 2018 study resulted in three recommendations, which I have restated in lay terms:

1. Establish an electromagnetic spectrum (EMS) superiority directorate by having a dedicated electronic warfare officer on the Air Staff, which is one of the two statutorily designated headquarters staff of the U.S. Air Force, to oversee and manage enterprise-wide EMS priorities and investments.

2. Combine and consolidate its current use of EW, including significant capabilities in individual platforms such as the active electronically scanned array radars (where the beam is steered without moving the antenna), distributed jammers fielded on various platforms, and upgrades in jamming electronics, such as the Electronic Attack Pod Upgrade Program (pods that identify, locate and counter sophisticated EW threats). (For completeness, in 2019 the navy also sought to develop a low-band tactical radio-frequency jammer pod to foil enemy counter-stealth radar systems.)[30]

3. Rebuild the EMS warfare warrior culture and expertise.[31]

The report admits that the air force, along with other services, took its eye off the electronic warfare ball, but the service still has a significant EW capability. Now the task is to integrate it against peer adversaries and have the warrior culture and expertise to use it.

Summary

Before leaving this chapter, I want to point out a unique feature of projected-energy warfare—namely, its cat-and-mouse nature. Nations prefer having dominance in directed energy; it positions them as the cat. When faced with superior directed-energy weapons, nations seek countermeasures, making them the mouse. For example, when discussing that China appears to be developing countermeasures against U.S. directed-energy weapons, the United States is the cat, and China, the mouse. The reverse holds true with Russia, as the United States as faces a superior Russian EW capability.

While China appears to be pursuing countermeasures as a defensive long-term, cost-effective strategy, the United States is pursuing them as an offensive strategy. In this context, you might conclude that China is content with being the mouse, while the United States is not. However, going back to China's cultural heritage and national goals, its philosophy aligns with an adaptation of Sun Tzu's advice: if the enemy has superior strength, evade him by using countermeasures. China's military goal is to dominate the Asia-Pacific region, while the United States has ambitions to be the dominant military on the planet. Thus, the United States pursues countermeasures that lend themselves to offensive electronic warfare.

While a cat-and-mouse strategy is not new in warfare, I think it will become more important as directed-energy weapons become dominant. Deception and confusion by the cat will be more critical than bombs and bullets against the mouse. Destroying a nation with cyber weapons could be the ultimate deception, as a covert cyberattack on a nation's electrical grids and financial institutions will almost assuredly be as devastating as a nuclear attack. Electronic warfare—such as jamming communications, radar, and GPS, and controlling the electromagnetic spectrum—will cause an adversary to lose situational awareness, thereby creating the ultimate confusion. Conversely, countermeasures by the mouse will become a necessity. All nations, especially technologically advanced ones such as the United States must use countermeasures to guard their computer networks against cyberattacks and to harden their electrical and electronic systems against directed-energy attacks.

In today's modern battlefield, a nation finds itself being a cat in one element of warfare and a mouse in another. If you are a mouse, nothing will serve you better than impenetrable armor. In earlier times, nations used walls, steel plates, and fortified defenses for armor. Today, the walls, steel, and fortifications are becoming obsolete as armor becomes invisible and more effective, bringing us to chapter 9, "Force Fields."

9

Force Fields

The good fighters of old first put themselves beyond the possibility of defeat, and then waited for an opportunity of defeating the enemy.

—SUN TZU

Much of what you have read in this book about directed-energy weapons had its roots in science fiction before it became science fact. One interesting aspect of good science fiction is its uncanny ability to predict future technology. Science fiction authors such as H. G. Wells envisioned laser weapons over a century before their invention. Another pillar of modern science fiction is Gene Roddenberry, the creator of the original *Star Trek* television series, who not only wrote about laser-like weapons but also introduced a new concept of energy force fields, or in the words of the *Star Trek* crew, "shields."

Roddenberry's force fields are invisible energy shields. For example, the fictional *Star Trek* force field is an energy barrier with varying degrees of strength, expressed in percentages, with 100 percent being the strongest. As *Star Trek* fans know, an adversary's repeated weapon strikes would cause the force field to weaken, potentially leaving the starship *Enterprise* vulnerable.

In one sense, I think Wells, Roddenberry, and other authors were science visionaries, giving us a glimpse of the future. Their visions took root in the popular culture of their time, inspiring engineers to develop the scientific capabilities found in their work. It is a case of life imitating art.

Previously, we discussed that all weapons use energy to wield destruction. In this chapter, energy plays a role in the development of real-life force fields that protect against matter, including shock waves, and those that protect against radiation. However, unlike the laser-like weapons found in *War of the Worlds* and *Star Trek*, force fields are still more in the realm of science fiction than science fact. Nonetheless, you will find their rudimentary beginnings as well as the challenges to their creation in this chapter.

Force Fields That Protect against Matter

Boeing's Force Field Patent

I am starting with Boeing's 2015 patent for a "method and system for shock wave attenuation via electromagnetic arc" because, more than any other, it closely resembles the force fields found in science fiction.[1] Additionally, shock waves from explosive devices are proving deadly, and Boeing's patent aims to protect our troops from them.

Improvised explosive devices (IEDs) caused over half of all American fatalities in the Iraq and Afghanistan Wars—more than thirty-five hundred in total—and wounded ten times as many. Although invisible, shock waves (i.e., sharp changes in pressure) were responsible for many of the fatalities and wounds. According to *Medium*, an online publishing platform, "When people are hit by the shock wave [caused by an IED], the sudden increase in air pressure can not only knock them down but can cause massive internal injuries. As little at [*sic*] 5 psi [pounds per square inch] over-pressure can rupture eardrums, and a 15 psi increase will begin to cause lung damage. An over-pressure of 35–45 psi has been shown to cause death in 1% of people, where an over-pressure between 55–65 psi will result in fatalities in 99% of people."[2]

Unfortunately, barriers, such as armored vehicles, that protect humans from blast debris do not always protect them from shock waves. Boeing's patent, if operational, will protect military vehicles, as well as their operators and passengers, from shock waves.

In one proposed configuration, Boeing's system uses a sensor to detect the explosion that produces the shock wave as well as its direction and distance from the military vehicle. Next, the engineers envision using a laser, microwave, or electrical arc generator to produce a plasma shield to protect "friendly assets," or the soldiers inside the vehicle. Boeing postulates the shock wave will be "attenuated in energy density by at least one of reflection, refraction, absorption, momentum transfer and magnetic induction."[3] This last statement may sound technically complicated, but there is an easy way to understand how the plasma shield works. You can think of it as a collection of ionized air molecules that is denser than air since it contains more particles in the form of ions and electrons. The density of the plasma blocks some of the shock waves. Sensing the explosion and creating the plasma have to happen in a fraction of a second, however, before the arrival of the shock wave.

Two elements of Boeing's patent are intriguing. First, it uses plasma, which is a hot ionized gas, the fourth state of matter. We see plasma every day but do not recognize it. The sun is plasma, only hot ionized gas. Other everyday uses of plasma are in fluorescent light bulbs and neon signs, which are tubes filled with gas. Plasma is the most common form of matter in the universe.

During my early work in the integrated circuit industry, we used plasma to deposit metals, which we termed "sputtering," as opposed to using evaporation. To form the plasma, we introduced a low-pressure inert gas, such as argon, in a vacuum chamber. Two oppositely charged plates would form and contain the plasma. In other words, they would strip electrons from the argon atoms, which then became positively charged ions and had a faint blue-green glow. The positively charged argon ions would race toward the negatively charged target of the metal we wanted to deposit. As the ions struck the metal, they dislodged or sputtered its atoms, depositing them on an integrated circuit substrate. The main benefit of this deposition technology was achieving a uniform deposit. In this process, the plasma was visible. Once formed, it glowed inside the deposition chamber. This observation suggests that the plasma shield envisioned by the Boeing patent would also likely be noticeable, given that its patent forms the plasma from the atmosphere by exciting it with an energy source.

Second, Boeing's patent lays claim to the use of plasma as a shielding methodology. Therefore, this patent would cover any defensive weapon that uses plasma as a shield, regardless of the means to achieve it or its configuration. A company that wants to produce such a weapon would have to obtain Boeing's permission and would likely end up paying Boeing a royalty. I have several patents. Some patents were operational inventions, and others were conceptual, meaning I only delineated potential methods to build them. In that sense, they exist only in my mind, but any company that manufactures them would be violating my patent. I suspect Boeing's patent may have elements of both, with some aspects being built and others being conceptual. We already know that a laser can ionize the atmosphere.[4] We do not know, however, how well this work may support Boeing's claims that its system will neutralize a shock wave. While the U.S. Patent Office considers the plasma shield patent novel and potentially buildable, in my opinion, building it is going to be a Herculean task.

When will we see a prototype of this energy shield? I think we are looking at several years to develop a prototype and another half a decade before it becomes operational. These projections are based on the following challenges its builders will need to overcome:

1. Sensors need to be able to detect a blast. This detection sensor seems the most achievable aspect of the patent and already exists in some weapons systems, which are discussed in the next section. Detecting a blast is relatively easy for modern sensors coupled with computer technology. For example, some explosions begin with a flash of light. Since light travels 900,000 times faster than sound (the shock wave is sound), the sensors will detect not only the explosion but also its direction and distance long before it reaches the vehicle.

2. Lasers, microwaves, or electrical arc generators need to become powerful, portable, and capable of generating the plasma from

the air between the vehicle and the blast. This aspect of the patent appears the most difficult, to my mind.

3. A weapon system needs to integrate the sensors and generate the plasma before the arrival of the shock waves. I think this aspect of the patent already exists in some weapons systems and is discussed in the next section.

Even if the three items were suddenly in place, we still do not know how effective plasma would be in shielding humans from shock waves. However, since Boeing is one of the giants in the U.S. military-industrial complex, it likely will be able to attract DOD funding to research and prototype the patent.

Active Protection System (APS)

An active protection system uses an explosion as well as the fragment action against an incoming kinetic weapon, such as a rocket-propelled grenade.[5] Here is how it works:

Sensors detect the incoming kinetic weapon.

A rocket or shotgun-like blast is directed at the threat to disrupt its effectiveness by defeating or deflecting it.

In a sense, you can think of it as a missile defense system applied to armored vehicles, such as tanks. Following are some of the active protective systems used by Russia, Israel, and the United States.

Russia: The application dates to the Drozd, a Soviet active protective system from the late 1970s, that was installed on T-55 tanks during the Soviet-Afghan War.[6] Since then, the system has gone through various upgrades. One current enhancement, Arena, designed in 1993, uses Doppler radar to detect incoming warheads and fires a rocket that detonates near the inbound threat to destroy it.[7] More recently, Russia deployed the Afganit APS on its next-generation T-14 Armata tank. The Afganit APS boasts an improved radar relative to the Arena's and an ability to defend against depleted uranium projectiles.[8] I am using the word "boasts" intentionally, as the information comes from the Russian news agency TASS and might be an exaggeration; however, if correct, it would mark a significant leap in tank defense. The U.S. Army uses alloys manufactured from depleted uranium because they make a warhead capable of penetrating an enemy tank's armor. A working Afganit would remove this vulnerability for Russian tanks.

Israel: Israel's two active anti-tank missile interceptor systems, Iron Fist and Trophy, began development in 2006 and 2007, respectively.[9] Both appear to work similarly, using radar to detect the threat and an explosive blast to neutralize it. However, according to *Israel Defense*, "Israel's Ministry of Defense intends to force Rafael Advanced Defense Systems and Israel Military Industries (IMI) to combine their active defence systems. Elta Systems will produce the radar, IMI the interceptor, and Rafael the command and control (C2) system."[10]

United States: The U.S. Army is currently buying the latest Trophy system for its M1 Abrams heavy tanks.[11] It is looking for a similar system for its 8x8 Stryker armored vehicles but is having difficulty finding a suitable one.[12] The army is not stating the specific problem it is encountering, but I suspect it has to do with the explosion of an ordinance near the vehicle. While a tank can handle that challenge and protect soldiers from the blast debris, the Stryker might not be able to sufficiently protect its human occupants from the shock waves.

I recognize that current active protective systems are not force fields in the sense of an invisible energy shield and rely on blasts or projectiles to destroy incoming threats. Because they are using energy to protect military vehicles from missiles and rockets, I think we can consider them an energy shield. Unfortunately, my research indicates they are not 100 percent effective; however, they do improve the survivability of the vehicles they defend. Therefore, the U.S. Army and other militaries are being pragmatic and using existing technology to address current threats.

Force Fields That Protect against Radiation

According to *Lexico*, powered by Oxford University Press, radiation is "the emission of energy as electromagnetic waves or as moving subatomic particles, especially high-energy particles which cause ionization."[13] Not many people give much thought to radiation. They never ponder its effects because radiation is common in our everyday lives, with radio waves used for wireless communication, microwaves to heat food, and X-rays to diagnose medical issues. Its commonality and lack of apparent effects on humans have led us to ignore it.

In addition to human-created radiation, naturally occurring sources include radioactive elements found on the earth, radiation trapped in the earth's magnetic field, and radiation from space. Most people show little concern for radiation from these sources as well.

The reality is that most of us go through life oblivious to radiation from any source since our contact with it is minimal. However, if you are an astronaut or a crew member of the International Space Station, then radiation takes on a different meaning. Once in space, exposure to radiation can become harmful and even lethal.

There are three kinds of space radiation:

1. Charged particles—ions (atoms or molecules with a net electric charge) and electrons (negatively charged subatomic particles)— trapped in the earth's magnetic field.

2. Electromagnetic radiation from the entire electromagnetic spectrum, from low-energy radio waves to high-energy (i.e., moving near the speed of light) gamma rays, that is emitted during solar flares.

3. Cosmic rays comprising high-energy protons, positively charged subatomic particles, and heavy ions (positively charged atoms or molecules stripped of one or more electrons) from outside our solar system.[14]

These kinds of space radiation represent ionizing radiation, which makes it capable of removing an electron from its orbit if it contacts an atom or molecule. This contact creates a positively charged entity, or an ion, and it gives this type of radiation its name. Examples of ionizing radiation include alpha particles, which are helium atom nuclei moving at high speeds; beta particles, or high-speed electrons; and gamma and X-rays, or high-energy electromagnetic radiation.

You can think of ionizing radiation as an atomic-scale bullet. On Earth, the atmosphere protects us from almost all harmful radiation, as virtually all cosmic rays collide with the oxygen and nitrogen of our atmosphere before they reach us. These same elements also absorb high-energy X-rays and gamma rays. Earth's magnetosphere protects us from the sun's solar flares, which are bursts of plasma traveling at nearly the speed of light. The sun continually bombards Earth in ultraviolet radiation, which is outside the visible light spectrum of humans, but the atmosphere's ozone layer absorbs much of this radiation. Even so, short periods of exposure to it results in getting a sunburn. Extended exposure can result in skin cancer. All life on Earth depends on its atmosphere and magnetic fields for protection against hostile radiation.

Space has no protective atmosphere or magnetic field. Ionizing radiation can damage whatever it hits. The most concerning are cosmic rays and short wavelength electromagnetic radiation, such as gamma rays. This form of ionizing radiation can damage integrated circuit electronics as well as humans. Unfortunately, it can easily penetrate the hull of a spacecraft and the skin of humans. The long-term effects of ionizing radiation are why NASA usually limits the cumulative time humans can spend in space to six months or less, depending on factors such as their altitude above Earth and exposure to solar flares. According to NASA, "Virtually any cell in the body is susceptible to radiation damage. The HRP [NASA's Human Research Program] is concerned with long-term health consequences of radiation exposure such as cancer, as well as adverse effects to the central nervous and cardiovascular systems."[15]

For extended space missions to be feasible, NASA needs to develop a better shield against space radiation. Today, NASA uses layered aluminum or titanium to shield its crews from radiation, since the dangerous particles are likely to deposit their energy within the shield. However, relying on added materials as a radiation shield soon becomes unrealistic, since more mass makes the spacecraft harder to launch.

In 2014 scientists with the European Organization for Nuclear Research (CERN) achieved a world-record current in a twenty-meter-long transmission line made of magnesium diboride superconductor.[16] While this has enormous implications for electrical power transmission, it also implies the creation of a massive magnetic field. Thus, scientists at CERN began collaborating with the European Space Radiation Superconducting Shield (SR2S) project.[17] Their objective is to develop a superconducting magnetic shield that can protect a spacecraft and its occupants from cosmic rays. According to Science Alert, "Previous announcements from SR2S have suggested that the issue of cosmic radiation can be solved within the next three years, with magnetic field technology enabling safe long-duration stays in space without harm from cosmic rays. The aim of the SR2S project is to create a magnetic field 3,000 times stronger than Earth's own magnetic field, with a 10-metre diameter protecting astronauts within or directly outside a spacecraft."[18]

This type of magnetic field could make deep space missions possible, but many challenges must be overcome, including testing the possible magnetic configurations as well as the magnetic field's effects on humans and electronics. We know Earth's magnetic fields protect us from space radiation, and that provides a significant incentive to develop a similar magnetic field for spacecraft that will protect humans as they traverse the solar system.

In the Next Decade

Although energy force fields are a staple of science fiction, we are a long way from developing them for warfare and space travel. I will leave you with two predictions.

First, I think the Boeing patent for "shock wave attenuation via electromagnetic arc" is the closest to a science fiction–like energy force field, given the "shield" is plasma. As such, it is denser than the atmosphere because it contains more particles, including ions and electrons. I expect Boeing to develop it and the U.S. military to deploy it within a decade.

Second, I think within a decade we will learn to create magnetic fields for spacecraft to shield its occupants from space radiation. This breakthrough will bring us closer to deep space exploration, possibly a permanent moon base, and a round-trip excursion to Mars.

The Problem with Creating Force Fields

This chapter is short for a reason: we still do not have a workable idea on how to create a *Star Trek*–like energy force field.

When I was a college student studying physics, I learned that there are four interactions, which are known as the fundamental forces: the gravitational and electromagnetic interactions, which produce long-range forces whose effects we observe in everyday life; and the strong and weak interactions, which produce forces at subatomic distances that govern nuclear interactions.[19] Like all students majoring in physics, I committed them to memory. However, something was missing. We understood them conceptually, but except for electromagnetic interactions, we were unable to create or control these forces. Mathematically, there was no single theory to explain them. Today that is still true.

Thus, today's scientists are finding creating a force field difficult. They are pursuing various theories in the hope that, eventually, they will develop a theory that explains all the fundamental forces. Perhaps then we will have a way to control the fundamental forces and create a real *Star Trek*–like force field.

Now, I would like to invite you to examine the future of directed-energy weapons in part 4, "The Coming New Reality."

FOUR

THE COMING NEW REALITY

10

Autonomous Directed-Energy Weapons

As seen across many nations, the development in autonomous
weapons systems is progressing rapidly, and this increase in the
weaponization of artificial intelligence seems to have become
a highly destabilizing development.

—JAYSHREE PANDYA

I n their televised introduction during Operation Desert Storm
in 1991, smart bombs made one thing clear: artificial intelligence
was going to change the nature of warfare. Millions around the
world watched in awe as the U.S. Air Force used them with surgi-
cal precision to neutralize military targets and minimize collateral
damage.

Along with many of my colleagues who helped build AI into weapons, I was as amazed as the public. The AI electronics we had talked
about with the Department of Defense and manufactured in our
facilities came to fruition. Seeing it all work was awe-inspiring, but
knowing people were dying as a result was horrific. In my mind,
though, my duty as a technologist was to ensure the U.S. military
had the types of weapons needed to prevail in combat. Sun Tzu's
quote from *The Art of War* expresses my feelings: "The art of war
teaches us to rely not on the likelihood of the enemy's not coming,
but on our own readiness to receive him; not on the chance of his
not attacking, but rather on the fact that we have made our position unassailable."[1]

Today we view the AI of the smart bombs used during Operation Desert Storm as rudimentary. They were guided munitions, meaning their programming enabled them to follow a laser beam. They were not autonomous weapons systems.

What Are Autonomous Weapons Systems?

The Department of Defense defines an autonomous weapon system (AWS) as "a weapon system that, once activated, can select and engage targets without further intervention by a human operator."[2] This definition centers on what autonomous weapon systems do, not on how they operate. Knowing how they function, however, reveals a crucial insight: according to the Congressional Research Service, an autonomous weapon system is "a special class of weapon systems that use sensor suites and computer algorithms to independently identify a target and employ an onboard weapon system to engage and destroy the target without manual human control of the system."[3]

This explanation of how autonomous weapons work also emphasizes their limitations—namely, the "computer algorithms" (i.e., artificial intelligence technology) that guide them also hinder them. This raises an important question: How good is today's artificial intelligence?

The Current State of Artificial Intelligence Technology

Today's AI technology falls far short of human intelligence. Specific endeavors that abide by rules, strategies, and preferred end points are now programmable into an "intelligent agent," or AI technology that can perform as well as an expert in a narrow field—for example, in the game of chess. However, checkers would bewilder a chess-playing intelligent agent because it does not have "general" artificial intelligence, meaning its technology does not replicate human intelligence. Every autonomous weapon that is in development or deployment has an intelligent agent at its core.

Intelligent agents follow set rules and strategies to achieve desired end points. They do not exercise judgment. Their inability to make judgments is the first major issue with autonomous weapons. This lack of human understanding could cause autonomous weapons to enter unintended engagements. Computer scientists are attempting to get around this deficiency by enabling machine learning, a technique that allows a machine to learn from experience and better perform a specific task. Even so, those machines still fall short of human intelligence. While they learn to do something, they do not understand it. For example, if you give a ten-year-old child a dozen pictures and request the child to pick those with a dog in it, the child can do it. With machine learning, a computer would eventually be able to replicate that feat. However, the ten-year-old does it by understanding what a dog is. The self-learning machine does it by trial and error. Given enough tests, it eliminates the errors. In the end, though, it still does not know what a dog is.

When will we have computers that can replicate human intelligence? In 2014 Vincent Müller and Nick Bostrom, professors of philosophy at Eindhoven University and Oxford University, respectively, published the paper "Future Progress in Artificial Intelligence: A Survey of Expert Opinion." In it, they reported, "The median estimate of respondents was for a one in two chance that high level machine intelligence [equivalent to a human's] will be developed around 2040-2050, rising to a nine in ten chance by 2075."[4] The implication is that most experts think it will take decades before AI technology achieves human intelligence. Without it, I think it would be fair to say AI technology would also lack human judgment.

This technology also poses another problem. Some current AI programs make trivial mistakes that could mean the difference between life and death in combat. For example, in 2018, a smart camera using facial recognition techniques to catch jaywalkers at intersections in Ningbo, China, misrecognized a photo of Chinese billionaire Mingzhu Dong on a passing bus ad. As a result, it flagged her as a jaywalker. The mistake went viral on Chinese social media. Although this mistake was trivial, what would happen if a similar misidentification occurs in combat? Would an autonomous weapon kill an ally?

The current state of AI is that it can perform as well as an expert in a well-defined field. However, even AI programs that learn from experience do not replicate human judgment, nor are they able to translate that learning from one endeavor to another. Thus, an intelligent agent chess-playing program may improve by playing additional games but will remain bewildered by checkers.

Autonomous Weapons Today

While some autonomous weapons exist today, they are the exception. One example is the U.S. Navy's Phalanx close-in weapon system for defense against anti-ship missiles. It is autonomous and has been in continuous deployment since 1973.

Although autonomous weapons are the exception, the situation is changing, with many autonomous weapons in development. We are at an inflection point: AI technology is reshaping the next generation of armaments that, in turn, will change the way we fight wars. However, before we discuss autonomous directed-energy weapons, I think it is best to clarify their associated ethical and political realities.

According to Future of Life, a volunteer-run research and outreach organization working to mitigate the existential risks facing humanity:

> Over the last few years, delegates at the United Nations have debated whether to consider banning killer robots, more formally known as lethal autonomous weapons systems (LAWS). This week [September 4, 2018] delegates met again to consider whether more meetings next year could lead to something more tangible—a political declaration or an outright ban. [As of this writing, no agreement exists regarding autonomous weapons.]
>
> Meanwhile, those who would actually be responsible for designing LAWS—the AI and robotics researchers and developers—have spent these years calling on the UN to negotiate a treaty banning LAWS. More specifically, nearly 4,000 AI and robotics researchers called for a ban on LAWS in 2015; in 2017, 137 CEOs of AI companies asked the UN to ban LAWS; and in 2018, 240 AI-related organizations and nearly 3,100 individuals took that call a step further and pledged not to be involved in LAWS development.[5]

These actions indicate many people around the world are concerned about building lethal autonomous weapons, which some term killer robots. This public outcry is a crucial point, and we need to understand the fundamental ethical question involved.

The International Committee of the Red Cross frames the moral question as follows: "The fundamental ethical question is whether the principles of humanity and the dictates of the public conscience can allow human decision-making on the use of force to be effectively substituted with computer-controlled processes, and life-and-death decisions to be ceded to machines."[6]

This ethical concern, given the current state of AI technology, is valid. However, to address it, we need to gain a more in-depth perspective because the United States and its closely pacing potential adversaries are building autonomous weapons. For now, though, in light of the ethical issues, the United States intends to use them semi-autonomously. The Russians, conversely, plan to use them autonomously. This difference has to do with control.

Autonomous versus Semiautonomous Lethal Weapons

Autonomous lethal weapons use AI technology to search, detect, and destroy enemy targets; therefore, AI technology makes the final decision regarding taking a human life. Semiautonomous lethal weapons work identically to a point but require a human operator to remain "in the loop"—that is, to make the final decision concerning taking human life—or "on the loop," meaning to supervise the weapon, presumably with the ability to abort the mission. In both cases, using semiautonomous weapons injects human judgment, which helps ensure their ethical use and avoid unintentional conflicts.

Department of Defense Policy

Department of Defense Directive 3000.09, "Autonomy in Weapon Systems," establishes the exercise of human wisdom during the use of autonomous and semiautonomous lethal weapons. It requires that "autonomous and semi-autonomous weapon systems shall be designed to allow commanders and operators to exercise appropriate levels of human judgment over the use of force."[7]

I recognize that the policy states "autonomous and semiautonomous," which may be confusing. Some weapon designs are only semiautonomous, meaning they always require a human commander to initiate an action that would result in the loss of human lives. Their design precludes operating them in an autonomous mode. Other weapon designs are autonomous, meaning the weapon can act on its own. However, as DOD Directive 3000.09 requires human control, the U.S. military uses an autonomous weapon in a semiautonomous mode. Let us consider an example.

Northrop Grumman X-47B

The X-47B is an experimental unmanned combat aerial vehicle designed by Northrop Grumman. An autonomous jet-powered stealth drone intended for aircraft carrier-based operations, it can take off and land on a U.S. aircraft carrier without human assistance. It is also capable of aerial refueling and acting on its own to search, detect, and destroy enemy targets.[8] In addition, the X-47B's stealth design allows it to penetrate highly contested enemy airspace.

In May 2015 the U.S. Navy announced the aircraft's primary test program complete and that the X-47B was successfully integrated into carrier operations. The navy plans to deploy four of these drones by the early 2020s.[9]

The X-47B and the drones deployed in the 2020s will be autonomous, but the navy intends to use them in a semiautonomous mode. In general, the U.S. military is concerned about taking human judgment out of the decision to use deadly force. The artificial intelligence of the X-47B does not rise to the level of human intelligence and, left to its own, may engage in unintended deadly force. Thus, the X-47B is an example of an autonomous weapon used in a semiautonomous mode. You may have a question at this point: Why make it autonomous if the intent is to use it semiautonomously?

If the navy needs the X-47B to penetrate highly contested airspace, using it semiautonomously may not be feasible, as that operation would require electronic communications between the operator and the drone. Such interactions would compromise the drone's stealth capabilities and alert the enemy to its presence. Thus, in combat, the U.S. military may need to violate Directive 3000.09 and use the x-47B in an autonomous mode.

Ethical Dilemmas

Given a choice, we should not allow a machine to make life-and-death decisions, regardless of its artificial intelligence. However, will we have an option during a war? As noted in the example of the x-47B, it will only remain stealthy if we do not communicate with it. Therefore, in extreme circumstances, such as in a highly contested airspace, we may need to allow the weapon to act autonomously.

For that reason, the DOD Directive 3000.09 permits exceptions regarding the use of autonomous weapons. For example, it explicitly excludes cyberspace weapons: the directive "does not apply to autonomous or semi-autonomous cyberspace systems for cyberspace operations."[10]

Why does it allow this exception? In part, it has to do with human reaction time because cyberspace weapons attack at the speed of light. In such an attack against a critical security network, for example, humans could not react with sufficient speed to inject control. However, I think there is another, subtler reason for this exception. I suspect U.S. military policy does not consider cyberspace weapons as having the same lethality as missiles, bombs, and bullets.

The reality, though, is that cyberspace weapons can be as deadly as nuclear weapons. If you doubt this, consider the consequences of a cyberspace weapon's destroying the electrical grids of a nation. As discussed in chapter 6, such an attack could lead to the death of millions.

Elements of Directive 3000.09 can be also be waived if approved by two undersecretaries of defense and the chair of the Joint Chiefs of Staff. Why are there waivers? War, by its nature, leads to unanticipated and extreme circumstances. There is no way to ensure any policy can anticipate all eventualities. Therefore, the Department of Defense requires a way to adapt to rapid changes in conditions during a conflict.

Why Build Autonomous Directed-Energy Weapons?

The reasons for building autonomous directed-energy weapons are identical to those regarding other autonomous weapons. According to *Military Review*, the professional journal of the U.S. Army, "First, autonomous weapons systems act as a force multiplier. That is, fewer warfighters are needed for a given mission, and the efficacy of each warfighter is greater. Next, advocates credit autonomous weapons systems with expanding the battlefield, allowing combat to reach into areas that were previously inaccessible. Finally, autonomous weapons systems can reduce casualties by removing human warfighters from dangerous missions."[11]

In addition to these points, I would add that autonomous directed-energy weapons, such as lasers, would enable the U.S. military to address swarm attacks by drones and missiles. For example, if a ship is under attack by a swarm of drones, it may run out of missiles and ammunition before stopping the attack. On September 14, 2019, Iran attacked Saudi oil-processing facilities using a swarm of seventeen drones.[12] Although the Saudis spent billions in past years on U.S.-made Patriot surface-to-air missiles, these weapons are only effective against high-flying targets such as enemy jets or ballistic missiles.[13] Another consideration is the cost of Patriot missiles. Each Patriot costs millions of dollars, whereas the value of the drones was less than $2,000 each. This discrepancy means the "cost to kill" ratio is not favorable. By contrast, a laser shot costs less than a dollar, and the U.S. Navy has already proven they are effective against drones.

Autonomous directed-energy weapons would be ideal for defense. For example, the U.S. Navy currently uses the autonomous Phalanx close-in weapon system as a last-ditch defense against anti-ship missiles. This weapon is similar to a machine gun and requires ammunition, which has to be stored. I think a laser weapon, which does not require ammo and can engage and destroy targets at the speed of light, could replace it. I also think a laser weapon would be more effective against drone swarm attacks than the Phalanx, since the speed of a laser beam simplifies targeting, thus enabling it to engage more targets at a time.

Overall, I think using autonomous weapons in a defensive role could be practical and would limit the possibility of unintended engagements. However, there may be circumstances that would require using autonomous directed-energy weapons offensively. For example, if adversaries deploy autonomous weapons, the U.S. military may also need to use them. Recall the adage "The best defense is a good offense." This wisdom echoes in many endeavors, but it is especially true in military combat. The reason maintaining a strong offensive position works so well in military conflicts is that it keeps the enemy continually engaged in defense, and nations do not win wars by defense.

What Autonomous Directed-Energy Weapons Make Sense

Given that current AI technology cannot replicate human judgment, I would limit the deployment of autonomous directed-energy weapons as follows.

Autonomous Cyberspace Weapons: Making cyberspace weapons autonomous is necessary since cyberattacks and electronic warfare attacks occur at the speed of light. There is no alternative. Involving a human would allow too much damage to occur before addressing it. This thinking aligns with the current DOD Directive 3000.09.

Autonomous Defensive Directed-Energy Weapons: In general, world militaries are scrambling to defend against the rise of cheap, low-tech threats such as drones, which are readily available on Amazon. Swarming tactics could conceivably overwhelm current defenses. Here again, autonomous laser weapons could be pivotal in defending against such attacks.

Autonomous Offensive Directed-Energy Weapons: I would limit the roles of these weapons to combat and after the start of hostilities. The most significant concern with offensive directed-energy weapons is the possibility of unintended engagements, especially those that could put human lives at risk. However, during combat, using offensive directed-energy weapons—such as autonomous cruise missiles packing a microwave punch that can take out enemy command-and-control centers—makes sense to me. For example, in situations where it is too risky to allow piloted stealth aircraft to penetrate highly contested airspace, I think we may need to use drones with appropriate directed-energy weapons for the mission.

Will the United States Deploy Autonomous Directed-Energy Weapons?

In my opinion, whether the United States would deploy autonomous directed-energy weapons depends on what weapons its potential adversaries use. In this context, I am referring to their using an autonomous weapon without human oversight.

Will potential adversaries of the United States use autonomous weapons? Unfortunately, the answer is yes. For example, in a 2017 United Nations meeting in Geneva, Russia announced that it would not adhere to an international ban, moratorium, or regulation on autonomous weapons.[14] Russia is sparsely populated and unable to match the military armies of countries such as the United States and China. For Russia, autonomous weapons represent a force multiplier. Russia also sells weapons around the world. If it does not have autonomy built into its next offerings, its clients may turn elsewhere.

If the Russians deploy autonomous weapons, how should the United States respond? It might not have any option but to use autonomous weapons also. It would be a "fight fire with fire" approach. Since the United States leads in most directed-energy weapons, exploiting that lead makes sense. From my viewpoint, when American lives are at stake, the U.S. military-industrial complex must provide the U.S. military with every possible advantage, including autonomous directed-energy weapons.

I acknowledge the ethical issues and the limitations of AI technology. Given both, I recommended only using offensive autonomous weapons after the onset of hostilities. War with autonomous weapons will be fraught with many unintentional casualties, but I would argue that is the nature of war. The United States has the right to defend itself, and it should not adhere to any policy that jeopardizes its survival.

Are Autonomous Directed-Energy Weapons Inevitable?

If we examine the x-47B, we can conclude that the United States is already building autonomous weapons. In time, such autonomous weapons will incorporate laser and microwave weapons for the reason that their use would enable the U.S. military to attack targets surgically and thus eliminate collateral damage. For example, a cruise missile with a microwave weapon would have the potential of disabling advanced weapons systems, such as Russia's long-range, surface-to-air defense system, the s-400.[15] The Russians claim their advanced surveillance radar and suite of missiles can track and target even U.S. stealth aircraft. However, according to the *National Interest*, their claim may not be credible. The *National Interest* reports, "While it is true that a layered and integrated air defense may effectively render large swaths of airspace too costly—in terms of men and materiel—to attack using conventional fourth generation warplanes such as the Boeing F/A-18E/F Super Hornet or Lockheed Martin F-16 Fighting Falcon, these systems have an Achilles' Heel. Russian air defenses will still struggle to effectively engage fifth-generation stealth aircraft such as the Lockheed Martin F-22 Raptor or F-35 Joint Strike Fighter."[16]

Nonetheless, as Russia improves the s-400 and eventually releases a next-generation s-500, penetrating highly contested airspace may prove too dangerous to allow human pilots to attempt the mission. In this extreme circumstance, we may see an autonomous drone similar to the x-47B used to destroy the s-500 installations.

Given these types of scenarios, I think the deployment of autonomous directed-energy weapons is inevitable, especially after the onset of hostilities. Even if the U.S. military lost several drones in such missions, it would be preferable to the loss of American pilots. It is far easier for a military commander to report equipment losses than human fatalities.

Another argument for the deployment of autonomous weapons is the increasing pace of warfare. Even human-crewed fighter planes must defend against computer-operated antiaircraft weapons systems. Imagine a swarm of missiles attacking a piloted U.S. fighter aircraft. Would it be feasible for the pilot to protect the aircraft against such an attack? Unfortunately, human reaction time is no match for computer processing speed. Therefore, I think even manned U.S. fighter aircraft will require autonomous weapons systems capable of engaging numerous enemy threats at once.

The Near Future of Warfare

The U.S. military is pushing to equip numerous platforms, such as ships, aircraft, and land-based units, with laser weapons within a decade. This push is fueled by two threats—missile and drone swarm attacks, and hypersonic missiles.

According to Next Big Future, a science and technology website, Undersecretary of Defense for Research and Engineering Michael Griffin told the Senate Armed Services Subcommittee on Emerging Threats and Capabilities in 2018, "We need to have 100-kilowatt-class weapons on Army theater vehicles. We need to have 300-kilowatt-class weapons on Air Force tankers. . . . We need to have megawatt-class directed energy weapons in space for space defense. These are things we can do over the next decade if we can maintain our focus."[17]

Scientists advised Undersecretary Griffin "that level of laser power [hundreds of kilowatts] is five to six years away, and a 'megawatt laser' is within a decade with persistent investment."[18]

As a result, the U.S. military is continually raising the budget for laser weapons, from about $300 million in 2018 to $1 billion in 2020. Lasers in the hundred-kilowatt range will address enemy drones, fast inshore attack craft, subsonic missiles, and ground-based air missile defenses, such as the Russian S-400. Megawatt lasers will address hypersonic missile threats and defend space assets.[19]

Undoubtedly the nature of warfare is changing, and its pace is increasing. The shifting nature and heightened speed of war are due to advances in military technology, such as artificial intelligence, which can enable swarm attacks, and hypersonic missiles, which travel five or more times faster than the speed of sound. To address these changes, the Department of Defense is emphasizing the development not only of directed-energy weapons, especially high-power lasers, but also of autonomous directed-energy weapons.

Concurrent with the evolution in the nature and pace of warfare, the modern battlefield is also expanding, moving beyond land, air, and sea to cyber and outer space. The United States and other nations historically used space assets to enhance their ability to fight terrestrial battles. In the parlance of the military, it is known as the militarization of space. Now, though, nations are racing to place weapons in space to deprive an adversary's use of its space assets. Placing armaments such as anti-satellite weapons in space is the weaponization of space, which takes us to chapter 11.

11

The Full-Scale Weaponization of Space

And there was war in heaven: Michael and his angels fought against
the dragon; and the dragon fought and his angels,

And prevailed not; neither was their place found any more in heaven.

—REVELATION 12:7–8, King James Version

I spent the better part of my career building the electronics for unique DOD space assets such as Milstar (Military Strategic and Tactical Relay), a constellation of five military communications satellites operated by the U.S. Air Force. Generally, a communications satellite relays television, radio, and telephone signals. While communications satellites are not new or unique per se, Milstar is. Let us briefly discuss its features.[1]

Geostationary Orbit: Milstar's orbit matches Earth's rotation; thus, to an observer on Earth, the Milstar satellite appears almost stationary. The Milstar satellite accomplishes this geostationary orbit by being approximately 22,200 miles above Earth's surface with an orbital speed of 7,000 miles per hour. This orbit is typical of communications satellites.

Why is this important? Since any one of the five Milstar satellites stays in almost the same position, communication with it is simplified. Antennas on the ground can be pointed at one of them and remain stationary. As a constellation of five satellites, Milstar also enables the U.S. military to have a global communications network. Last, satellites in geostationary orbit are more difficult to destroy by ground-based lasers and missiles. The critical phrase in the last sentence is "more difficult." Given that low-Earth orbit satellites are typically a few hundred miles above Earth, lasers can blind them, and other weapons can destroy them. Until recently, geosynchronous satellites, being over twenty thousand miles above Earth, were out of reach of both missiles and lasers, but that is no longer the case. According to the *Economist*, "Perhaps the simplest way to attack a satellite is to hit it with a missile from Earth. This is what China did in 2007, taking out one of its own weather satellites, and what India did this March [2019]. Such attacks are easier to do when the target is in a low orbit [as Russia also demonstrated in 2020]. But China has tested missiles apparently capable of getting all the way to geostationary orbit."[2]

The U.S. Combined Space Operations Center (CSPOC), in a windowless building located at Vandenberg Air Force Base, California, tracks just about everything in space, from satellites to space debris the size of softballs. The Eighteenth Space Control Squadron, within CSPOC, is tasked with managing "space situational awareness" and tracking any potential missile attack on a U.S. space asset.[3] Attacking a low-Earth orbit satellite via a missile would take time, typically measured in minutes. Attacking a geosynchronous orbit satellite would take hours, and only liquid-fueled missiles can reach its orbit. Such an attack would be relatively easy to detect because the United States has a classified number of Defense Support Program satellites in geosynchronous orbit designed to detect missile launches.[4] Once identified, the spectrum for retaliation is enormous.

As noted, high-power lasers also can blind a surveillance satellite and potentially destroy other low-Earth orbit satellites. As the laser beam travels at the speed of light, it would only take a fraction of a second to reach its target. The U.S. military has this type of high-power laser in its arsenal and leads the world in laser and microwave weapons. China and Russia are also developing high-power lasers for this task.

If U.S. space assets come under attack by any means, their defense and a counterattack would be under the purview of the U.S. Space

Command, a unified combatant command of the U.S. armed forces that is responsible for space warfare. Any attack on U.S. space assets would be considered an act of war. Based on the U.S. military's current capabilities, it is not apparent that such an attack would succeed. If it did, it would unleash, as Revelations 8:7 states, "hail and fire" against the suspected perpetrator.[5]

Nuclear Survivable: The high radiation associated with nuclear detonations will not affect Milstar due to its radiation-hardened (rad-hard) electronics. At Honeywell, I was responsible for this specific area and directed about five hundred engineers and technicians to develop and manufacture the rad-hard electronics for the satellites. Although their exact design and manufacture are secret, I can share some of the design and construction attributes that are in the public domain. Per Honeywell's "Radiation Hardened SOI CMOS Technology" brochure, "Honeywell has been at the forefront of radiation hardened integrated circuit (IC) technologies for decades."[6] The attributes of an SOI CMOS are low noise (i.e., low current leakage between transistor elements), low power consumption, and high-speed performance. The design and construction techniques allow the resultant electronics to withstand the constant bombardment of space radiation, typically generated by the sun, as well as an electromagnetic pulse caused by a nuclear weapon.

Secure Communication: Milstar satellites, according to the U.S. Air Force, "perform all communication processing and network routing onboard, thus eliminating the need for vulnerable land-based relay stations and reducing the chances of intercepted communications on the ground."[7] Milstar satellites are also resistant to enemy jamming. This capability has to do with the signal strength and the communication coding, which remain classified. I will say this, though: it is hard to jam a strong coded signal.

I explained the Milstar satellite system and its capabilities as an example of the militarization of outer space. In this context, "militarization" refers to how space assets enable the U.S. military to fight more effectively on Earth. The Milstar satellites are not weapons; they are communications satellites that the U.S. military uses to communicate globally. The United States began militarizing space in the early 1960s. An example that most people are familiar with is the Global Positioning System, whose satellites provide the location, velocity, and time synchronization for air, sea, and land travel.[8] Civilian vehicles, aircraft, and ships use G P S for navigation. The U.S. military uses it similarly, but the services also use it for unique military purposes, such as guiding a cruise missile to its target. The military considers G P S a dual-use technology.

U.S. Military Satellites

This section is a thumbnail sketch of the types of U.S. military satellites that relate to the militarization of space. They fall into four classifications:

1. Communications Satellites: The Milstar constellation of five military communications satellites is one example, whose primary function is to provide the U.S. military with a secure global communications network.

2. Navigational Satellites: The Global Positioning System's satellites, for example, enable civilian and military navigation, as well as military targeting for cruise missiles.

3. Reconnaissance Satellites: These satellites observe Earth for military purposes and use infrared sensors that track missile launches and "electronic sensors that eavesdrop on confidential conversations."

4. Weather Satellites: The U.S. military uses these satellites to forecast the weather, which can be crucial to ensure the safety of naval vessels, land forces, and air forces. Modern weather satellites can observe cloud systems, "sand and dust storms, auroras, city lights, snow cover, energy flows, fires," ice cover, "effects of pollution, and ocean currents."[9]

While there are other types of satellites, these four classes represent the U.S. military's primary space assets. The most important observation here, however, is that the satellites in these four categories are not weapons.

Another essential fact about the militarization of space is that about 95 percent of all satellites in space are dual use, meaning they can perform civil as well as military functions. GPS, for example, is a dual-use satellite system. Many observation satellites are also dual use since they can map terrain or military movements. Weather satellites are also dual use in that besides detecting weather patterns, they also track environmental data that affects communications and radar. Milstar satellites, though, are not a dual use, as they are only for secure global military communications.

The Weaponization of Space

The "weaponization of space" refers to placing weapons in outer space. As noted in chapter 1, currently the United States has formally agreed to one UN treaty about space, the Treaty on Principles Governing the Activities of States in the Exploration and Use of Outer Space, including the Moon and Other Celestial Bodies, which is colloquially known as the Outer Space Treaty. Russia, the United Kingdom, and the United States sponsored the treaty, which entered into force on October 10, 1967, and forms the basis of international space law. As of June 2019, 109 countries are parties to the treaty, meaning they have signed and ratified it. Another 23 have signed the agreement but have not ratified it.[10] Among them are the People's Republic of China and North Korea. (I am explicitly referring to the People's Republic of China here to differentiate it from the Republic of China, which we colloquially call Taiwan.)

Before we proceed further, we should understand the salient principles of the Outer Space Treaty:

It prohibits the placing of nuclear weapons in space.

It limits the use of the moon and all other celestial bodies to peaceful purposes.

It proclaims space shall be free for exploration and use by all nations.

It stipulates that no nation may claim sovereignty of outer space or of any celestial body.[11]

Given that neither the People's Republic of China nor North Korea has ratified the treaty, the agreement does not legally bind China and North Korea. Meanwhile, both China and North Korea have launched satellites, with China having 280 and North Korea 2.[12]

North Korea launched its first orbiting satellite on December 12, 2012; it is in low-Earth orbit.[13] Although this satellite obtained orbit, it is tumbling and appears uncontrolled. North Korea launched its second satellite on February 7, 2016, after conducting a nuclear test on January 6, 2016. It is crucial to understand that North Korea had nuclear weapons since 2006, long before launching its satellites, which are in low-Earth orbit.[14] While North Korea states its satellites are for surveillance, experts detect no signals from them.[15] The second satellite initially tumbled in orbit, but North Korea corrected that, meaning it can communicate with the satellite. The question that concerns me is whether either satellite contains a nuclear weapon. Since the satellites are in low-Earth orbit, the detonation of a nuclear weapon would create a high-altitude electromagnetic pulse that could severely cripple any country within the detonation zone, including the United States.

Over the last few years, we have watched North Korea launch missiles to achieve altitude, not range. Although some consider such launches a failure, they may not be. According to the Observer Research Foundation, a respected think tank based in India, "Dr. Peter Vincent Pry, executive director of the Task Force on National and Homeland Security and chief of staff of the Congressional EMP Commission, recently suggested that North Korea's so-called [failed] 'missile launch looks suspiciously like practice for an EMP attack. The missile was fired on a lofted trajectory, to maximize, not range, but climbing to high-altitude as quickly as possible, where it was successfully fused and detonated—testing everything but an actual nuclear warhead.'"[16]

Based on its current capabilities, including missiles, satellites, and nuclear weapons, North Korea appears to have the means to execute a HEMP attack, making it a threat to the United States and other nations.

Although the Outer Space Treaty prohibits placing weapons of mass destruction in space, it does not prevent the weaponization of space. Weapons such as guns, lasers, microwaves, and missiles with a conventional warhead are not weapons of mass destruction; therefore, their use in space does not violate the treaty.

Potential adversaries are aware of the United States' space assets and their criticality to its military. These adversaries, fearing the United States' capability to use space assets in combat against them, are focusing on depriving the U.S. military of their use.

According to the U.S. Defense Intelligence Agency, both China and Russia are weaponizing space, including laser weapons and ground-based anti-satellite missiles.[17] An excerpt from the agency's 2019 report notes the following challenges:

> Chinese and Russian military doctrines indicate that they view space as important to modern warfare and view counter space capabilities as a means to reduce U.S. and allied military effectiveness. Both reorganized their militaries in 2015, emphasizing the importance of space operations.
>
> Both have developed robust and capable space services, including space-based intelligence, surveillance, and reconnaissance. Moreover, they are making improvements to existing systems, including space launch vehicles and satellite navigation constellations. These capabilities provide their militaries with the ability to command and control their forces worldwide and with enhanced situational awareness, enabling them to monitor, track, and target U.S. and allied forces.
>
> Chinese and Russian space surveillance networks are capable of searching, tracking, and characterizing satellites in all earth orbits. This capability supports both space operations and counter space systems.
>
> Both states are developing jamming and cyberspace capabilities, directed-energy weapons, on-orbit capabilities, and ground-based antisatellite missiles that can achieve a range of reversible to non-reversible effects.[18]

China is second only to the United States in the number of operational satellites, with 280 versus the United States with 830.[19] Also, China is rapidly expanding its space assets, despite officially advocating the "peaceful use of space."[20]

China has an operational ground-based anti-satellite missile that can target space assets in low-Earth orbit. Additionally, the *National Interest* reported in 2018 that China has secretly tested anti-satellite weapons capable of attacking high Earth orbit satellites.[21] China is also planning to field a ground-based laser weapon that can blind low-orbit space-based sensors by 2020.[22]

Russia is lagging behind China in anti-satellite arms but is making progress. According to CNBC, "In October [2018], CNBC learned that a never-before-seen Russian missile was a mock-up of an anti-satellite weapon that will be ready for warfare by 2022. Images of the mysterious missile on a modified Russian MiG-31, a supersonic near-space interceptor, appeared in mid-September." CNBC also reports the Russian anti-satellite weapon, which Russia has successfully tested several times, is built to target "communication and imagery satellites in low Earth orbit."[23]

Although China and Russia are developing anti-satellite weapons capable of destroying U.S. satellites, such an attack would have two tragic consequences. First, it would be an act of war and would likely warrant retaliation. In the extreme case, the retaliation might include a nuclear strike. It is crucial to understand that U.S. nuclear-tipped intercontinental ballistic missiles do not require space assets to hit adversaries anywhere on Earth. Second, any kinetic attack against a space asset would leave a massive and dangerous debris field, which would render space exploration and satellite function impossible for everyone for decades. Any space debris, even the size of a U.S. quarter, would inflict disastrous damage to any spacecraft because of their kinetic energy, which is a function of their mass times their velocity squared. Debris in low-space orbit (i.e., between 150 to 300 miles above the earth) travel at about 17,500 miles per hour, or about seven times faster than a bullet, to maintain orbit against the earth's gravitational pull.[24] Therefore, you can think of space debris as speeding bullets and cannonballs. The U.S. Space Shuttle needed to replace its windows routinely due to damage by paint chips from the booster rockets. Larger debris, such as the booster rockets, are the size of a school bus. Any collision between larger debris and a spacecraft would be catastrophic.

Thus, both Russia and China are taking another approach to disable U.S. satellites. According to *5280*, a Denver-based magazine, "Russia has recently launched satellites with the ability to cozy up next to, listen to, observe, and (some have suggested) even grab on to other satellites. China and Russia both have been developing lasers that have the ability to 'blind' satellites by damaging their imaging sensors. . . . Perhaps the most concerning technique—one Russia has allegedly already put to use in the Baltics—is GPS 'spoofing,' or the ability to mimic or modify a GPS signal to confuse or misdirect an enemy."[25]

The weaponization of space by potential U.S. adversaries is causing considerable concern. This issue is not new. Let us digress to understand the events that made it a critical matter for U.S. military planners.

The United States Gets a Wake-up Call from Space

In January 2001 a congressional space commission headed by Donald Rumsfeld, a secretary of defense under President George W. Bush, recommended that the United States "develop and deploy the means to deter and defend against hostile acts directed at U.S. space assets and against the uses of space hostile to U.S. interests." The commission sought to avoid, in its own words, a "Space Pearl Harbor." The group unanimously recommended that the United States invest in technologies that would enable it to field space assets one generation ahead of "what is available commercially."[26] Its report also suggested that the U.S. government encourage its commercial space industry to field systems one generation ahead of its international competitors' systems as well.

What actions resulted from the 2001 commission remain unclear. Meanwhile, on January 17, 2007, *Aviation Week*, a weekly magazine on the aerospace, defense, and aviation industries, reported, "U.S. intelligence agencies believe China performed a successful anti-satellite weapons test at more than 500 mi. altitude Jan. 11 destroying an aging Chinese weather satellite target with a kinetic kill vehicle launched on board a ballistic missile."[27]

China's new anti-satellite weapon alarmed the U.S. military. Additionally, the destruction created a debris cloud of more than three thousand pieces, the largest ever. The debris will remain in orbit for decades, posing a significant collision threat to other space objects in low-Earth orbit. It was a reckless weaponization of space demonstration.

This anti-satellite weapon test was a wake-up call for the United States. Its military planners asked themselves one question: Does this anti-satellite weapon give China the potential to execute a Pearl Harbor–style attack in space?

A Space Pearl Harbor on U.S. Space Assets

A Space Pearl Harbor sneak attack would be difficult to execute. Let us understand why by examining four types of potential attacks on U.S. space assets.

A Missile Attack

Attacking U.S. space assets with missiles or anti-satellite weapons, similar to what China used to destroy its weather satellite, would be trackable via the U.S. Defense Support Program satellites and take minutes to hours to accomplish, depending on the orbit of the satellites under attack. The United States would consider such an attack an act of war, as it did when the Japanese attacked U.S. military assets at Pearl Harbor on December 7, 1941.

After confirming an adversary's intention to attack U.S. space assets, the U.S. military would launch a massive counterattack against that enemy and focus on the following targets:

Space Assets: China, Russia, and other potential adversaries have relatively few satellites compared to the United States as of 2020:

United States—1,327 satellites

China—363 satellites

Russia—169 satellites

North Korea—2 satellites

Iran—1 satellite[28]

The U.S. military has both missiles and lasers capable of destroying the entire space assets of its potential adversaries. For example, as early as 1997, the U.S. Army used a laser to destroy a low-Earth orbit satellite, and in 2008 the U.S. Navy carried out Operation Burnt Frost, firing a ship-launched missile to destroy a defunct U.S. spy satellite.[29] As noted previously, however, the United States, China, and Russia are reluctant to test anti-satellite weapons because destroying a satellite leaves a debris field, which causes a severe problem. The International Space Station already has to tweak its orbit routinely to avoid space junk.[30] Given the historical facts, we can reasonably infer the U.S. military is capable of destroying an adversary's space assets.

Missile Launch Sites: All of an adversary's missile sites, land or sea based, would become a primary target for two reasons: First, it would prevent the adversary from launching more missiles at U.S. space assets. Second, it would prevent an adversary's nuclear strike on the United States. Unfortunately, if events unfold to this point, the continued existence of humanity would be in doubt. Recall that Russia has nuclear parity with the United States, and many of the Russian nuclear-tipped missiles are on mobile launchers. In my opinion, destroying them all would be difficult, but failing to do so would ignite the kind of nuclear exchange feared during the Cold War. Even a small number of nuclear missiles reaching the U.S. and Russian homelands would likely lead to a nuclear winter and threaten human existence.

To a lesser extent, the same is true of China. A subsequent war between the United States and China also would likely end in an apocalyptic scenario.

Airpower Assets, Including Planes, Airfields, and Aircraft Carriers: The U.S. military's strategy is to gain air dominance in any conflict. Achieving that goal requires the destruction of an adversary's air force assets.

Strategic Command: Military command centers are legitimate and high-value targets. Removing them could destroy the adversary's ability to engage the U.S. military.

Leadership: The head of state of an adversary at war with the United States is a legitimate target. Removing the adversary's leader may be sufficient to cause the adversary to surrender. This strategy is, in military parlance, a "decapitation strategy." Eliminating an adversary's military and political leadership is a warfare strategy that dates back thousands of years to advocates such as Sun Tzu.[31]

In sum, any nation attempting to destroy U.S. space assets with missiles or anti-satellite weapons would invite a massive counterattack that could lead to a nuclear exchange and threaten human existence.

A Laser or Microwave Attack

Attacking U.S. space assets with a laser or microwave weapon is more problematic. Let us understand why.

Laser beams and microwave signals travel at the speed of light and would reach a satellite almost instantly. A laser can blind a surveillance satellite and a microwave can destroy its electronics without either attack leaving a debris field. If a laser destroys a satellite, however, it will leave a minimal debris field consisting of the plasma created by the laser as it tunnels into the spacecraft.

Pinpointing the origin of the laser or microwave generator, however, may be difficult. Several satellites could be lost before the U.S. military determines its location and the nation responsible for the attack. For example, an adversary may use more than one laser or microwave weapon, including a combination of them. The laser or microwave attack also may come from an adversary's satellite in orbit, making determining the locations more complex.

To understand the difference between a laser and a microwave weapon, see table 1, which Tom Wilson, a member of Rumsfeld's congressional space commission, presented in January 2001.[32]

Table 1. Comparison between high-power laser and high-power microwave anti-satellite weapons

High-power laser weapons	High-power microwave weapons
Irradiates a selected spot on a single target with high precision	Attacks area targets that may include groups of targets
Must precisely be aimed and pointed at susceptible area of target	Needs only be directed generally toward the intended target
Inflicts heavy damage on selected spot	Inflicts more subtle damage on electronic components
Will not operate through clouds	Largely unaffected by clouds
Heats, melts or vaporizes a selected spot which in turn destroys or disables the target	Generates high electric fields over the whole target which in turn disrupts or destroys vulnerable electronic components

Source: Tom Wilson, "Threats to United States Space Capabilities," Federation of American Scientists, 2001, https://fas.org/spp/eprint/article05.html#rft70.

Although identifying the source of a laser or a microwave attack on U.S. satellites may prove challenging, the United States has sufficient capability, in the form of redundant surveillance satellites and drones, to make that determination. At that point, the U.S. military would launch a massive counterattack along the lines already discussed.

A Cyberspace Attack

The most challenging attack to attribute to an adversary is a cyberspace attack. It could include electronic warfare—disrupting the electromagnetic field, for example, and damaging a satellite's communications—and cyber warfare, such as using computer technology to disrupt a satellite's function.

Thus, a cyberspace attack requires more sophistication to determine the perpetrator. Unfortunately, adversaries often hire proxies or hackers who are located in other countries and who use other countries' systems. Even when the United States identifies the perpetrator, proportional responses fall into a gray area. For example, Russia uses electronic warfare to jam GPS signals, a spoofing problem NATO forces first detected during their large-scale Trident Juncture exercises in Norway at the end of October 2018.[33] The U.S. Army responded by making the GPS signal jam-resistant and fielding next-generation weapons that use an inertial navigation system when GPS is entirely unreachable. Note that the U.S. approach to Russia's cyberspace attack on its GPS did not result in direct military action against Russia; instead, the United States deployed countermeasures.

A Particle Beam Attack

For completeness, I discuss here the potential of an enemy's using a particle beam weapon to destroy U.S. satellites. I did not mention it previously for two reasons:

1. Particle weapons project an intense beam of atomic or subatomic particles (i.e., elementary particles) that travel only a fraction of the speed of light, obtaining "relativistic velocities." In a sense, a particle weapon is a gun that shoots elementary particles, which can have a positive, negative, or neutral charge. With this understanding, I do not consider a particle weapon a directed-energy weapon but respect differing opinions.

2. Particle weapons are only useful in space. The atmosphere's density renders Earth-based particle weapons useless.

Although scientists use ground-based particle accelerators for high-energy nuclear physics research, no nation is fielding a particle beam weapon. The U.S. government has decided not to fund particle beam weapons because the technology is still in an embryonic stage. According to then undersecretary of defense for research and engineering Michael Griffin in 2019, "The funds will go instead toward more fundamental research aimed at making lasers more powerful."[34]

The United States is openly acknowledging its pursuit of lasers as a weapon for both terrestrial and space warfare. It remains silent, however, on killer satellites.

Killer satellites are orbiting anti-satellite weapons. While the United States has not publicly acknowledged having killer satellites in space, it does admit to having the Boeing X-37B, also known as the Orbital Test Vehicle. The X-37B is a reusable robotic spacecraft that resembles a miniature version (29 feet long) of NASA's 122-foot-long Space Shuttle, which retired from service in 2011.[35] The U.S. Air Force boosts the X-37B into space by a launch vehicle. When it obtains low-Earth orbit, it can maneuver in orbit. This capability makes the X-37B difficult for adversaries to track and enables it to approach other spacecraft in low-Earth orbit.

Similar to the Space Shuttle, the X-37B has a large internal payload compartment, with two payload doors that expose the chamber to space. The bay is large enough to hold satellites, missiles, nuclear weapons, a microwave weapon, or a laser. Whatever payload the X-37B carries is a closely guarded secret, as is its mission.

Since the X-37B has no crew, it can and has stayed in space for months to years. When the vehicle completes its mission, it reenters Earth's atmosphere at a speed of Mach 25 (i.e., twenty-five times the speed of sound) and, like the U.S. Space Shuttle, lands as a spaceplane.[36]

The U.S. Air Force has successfully orbited the X-37B five times, with the last orbit setting a new record. According to CNN, "An Air Force X-37B spaceplane just completed its 718th day in orbit [as of August 28, 2019], making it the most extended mission yet for a secretive military test program."[37] On October 27, 2019, the X-37B successfully landed at NASA's Kennedy Space Center in Florida, after spending 780 days in orbit.[38]

The U.S. Air Force has two X-37B spaceplanes, which appear more capable than any Russian or Chinese anti-satellite weapons. While the U.S. Air Force is not labeling the X-37B an anti-satellite weapon, it has all the capabilities required: First, it can move around in orbit and interact with other satellites. Next, with its large payload compartment, it could carry equipment to attach surveillance equip-

3. The x-37b Orbital Test Vehicle. Courtesy of U.S. Air Force.

ment to an adversary's satellite, disturb its alignment, jam its signals, destroy its electronics with microwaves, or even capture it and return it to the U.S. military for examination. Last, the x-37b would also be capable of attacking terrestrial targets by releasing missiles either while in orbit or after its reentry. Since it is traveling at twenty-five times the speed of sound, I doubt any nation could defend against such an attack.

Russia is also launching satellites that behave as killer satellites; that is, they can move around in orbit and interact with other satellites. These capabilities, which are similar to those of the x-37b, are suspicious since they could enable their spacecraft to interfere with other satellites. While a missile attack would be overt, one satellite disturbing another may appear accidental, depending on the circumstances. According to the *National Interest*, "In August 2018, U.S. official Yleem Poblete accused Russia at the Geneva Conference [on] Disarmament of deploying satellites into space designed to kill other satellites. Moscow, however, insists that these are simply inspector satellites designed to glide within close range of other satellites, diagnose problems with them and potentially even fix them."[39]

This Russian capability is troubling to the U.S. military, but space is still the Wild West. Space law is in its infancy. The Russians are taking advantage of this fact. Since 2013 they have launched four similar "inspector satellites." While their claim about these satellites may be valid, any satellite that can get close enough to repair a satellite can also compromise or destroy it. A satellite that can fix or destroy other spacecraft is another example of dual-use technology. Additionally, in 2020, according to Gen. John Raymond, Space Command's commander, Russia tested "a so-called 'direct-ascent anti-satellite' weapon," thus expanding its space arsenal.[40]

China is also developing killer satellites, but it likewise claims they are for inspection and repair purposes.[41] The Chinese also maintain these satellites will clean up space debris. Again, any satellite with these capabilities can be a dual-use killer satellite.

In one last observation, nations will only use killer satellites during a war since using them is an act of war. To my mind, space war is the next likely battlefield. At one time, nations fought all battles on land and at sea. With the invention of the airplane, hostilities moved to the skies. With the development of radio and the internet, cyberspace became a battlefield. Now, as more nations gain entry to space, it is only a matter of time before space becomes a battlefield. Therefore, the United States is preparing for the coming space wars.

U.S. Space War Strategy

On June 18, 2018, President Donald Trump directed the Pentagon to create a space force, and on December 20, 2019, he signed into law the National Defense Authorization Act for Fiscal Year 2020. "The bill creates the U.S. Space Force as the sixth branch of the U.S. armed forces," with a role to oversee missions and operations in the rapidly evolving space domain.[42] Previously, the U.S. Air Force had this mission. The creation of a U.S. Space Force is the first new military service in more than seventy years, since establishing the U.S. Air Force in 1947.[43] From my perspective, I think having a dedicated service branch for space makes sense, especially as Russia and China continue to weaponize space. The risks are enormous. Significant losses of U.S. space assets would cripple our global distribution networks for our food, energy, and transportation, as well as destroy global banking and communications.

How would the United States cope during a war without the use of its space assets? Would it be able to continue to use cruise missiles and advanced weapons, which rely on space assets? Alternatively, would the world's most advanced military find itself fighting in a World War II–like scenario?

These questions are troubling, but they raise critical military issues. Apparently the United States has thought them through and has developed a strategy in which it would use its drones, such as the Global Hawk, and its spaceplane, the X-37B, to patrol the skies. This strategy has the code name Triple Canopy, but the U.S. military does not acknowledge its existence.[44] However, those drones would perform surveillance and guide missiles. In my opinion, the loss of space assets would be significant, but the U.S. military would still be able to guide precision weapons and have situational awareness.

The real issue is not winning a war but ensuring the survival of humanity. Global networks, from food distribution to banking, rely on space assets. After losing our space assets, we would find ourselves plunged back into early twentieth-century life but without the systems and the skills that allowed us to survive during that time. To say it would be catastrophic would be a gross understatement.

While you may think that we eventually could replace the space assets with new versions, it is unlikely. The debris field left behind by a space war would render space unusable for decades and perhaps centuries.

The Thucydides Trap

Until now, humanity has avoided destroying itself, even though it has possessed the means since about the mid-twentieth century. We have sufficient nuclear weapons to destroy the world twice over, but we used judgment to avoid Armageddon. To be clear, we came close on several occasions, but human wisdom prevailed.

Now, we find ourselves at a tipping point. Will we be able to avoid the Thucydides Trap? Graham Allison proposed the theory of Thucydides Trap, postulating that war between a rising power and an established power is inevitable.[45] While Allison's postulate is an assumption, it does appear well rooted in history. According to Allison, "As China challenges America's predominance, misunderstandings about each other's actions and intentions could lead them into a deadly trap first identified by the ancient Greek historian Thucydides. As he explained, 'It was the rise of Athens and the fear that this instilled in Sparta that made war inevitable.' The past 500 years have seen 16 cases in which a rising power threatened to displace a ruling one. Twelve of these ended in war."[46]

Currently, China is a growing power, and the United States is at the top of the power pyramid. If Allison is correct, and China and the United States fall into the Thucydides Trap, I think the world as we know it will cease to exist. No humans will remain to claim victory. Although China has a smaller nuclear inventory, with about 320 nuclear weapons versus the United States' 5,800, I think a war with China could result in significant losses of space assets on both sides and a nuclear exchange.[47] Like Russia, China's missiles are on mobile launchers and would be difficult to destroy. Some would inevitably survive and could be used to launch their missiles. As discussed previously, even a limited nuclear exchange is likely to result in a nuclear winter and the type of destruction seen in apocalyptic movies.

Additionally, China and the United States could unleash their anti-satellite weapons and successfully destroy a portion of each other's space assets, which would leave a debris field that would prevent the use of space for decades to centuries. The destruction of space assets would also cause global distribution and financial transactions to cease, resulting in countless fatalities over time.

This concern brings us to chapter 12, "Not Gambling with the Fate of Humanity."

12

Not Gambling with the Fate of Humanity

New technology is not good or evil in and of itself.
It's all about how people choose to use it.
—DAVID WONG

We have come a long way, and our journey has revealed an arsenal of new directed-energy weapons. These weapons will force significant changes in military strategy and disruptions to existing treaties. Concurrent with the emergence of directed-energy weapons is the insertion of artificial intelligence in war armaments across the entire weapons' spectrum. Unfortunately, taken together, these emerging capabilities are increasing the potential for nations to engage in c-war, or war at the speed of light. Let us examine why.

Destabilizing the Doctrine of Mutually Assured Destruction

The doctrine of mutually assured destruction is not a formal treaty but exists in the realm of military policy and strategy.[1] Surprisingly, the doctrine of MAD has kept nuclear powers from engaging in a nuclear war. You may wonder why I used the word "surprisingly." The reason is MAD rests on an almost insane premise: a full-scale use of nuclear weapons by two adversaries would destroy both the attacker and the defender.[2] I use the word "insane" because the doctrine of MAD justifies having nuclear weapons as a deterrent to full-scale nuclear war, thereby justifying states to develop or maintain these weapons of mass destruction. To me, that is insane.

The underlying premise of MAD is a belief that small nuclear states, having fewer nuclear weapons, can deter aggression by large nuclear nations. Some believe that Iraq might have prevented the U.S. invasions in 1991 or 2001 had it possessed nuclear armaments.[3] This kind of thinking is why North Korea is developing nuclear weapons and a missile delivery system. It is ironic because nations believe the best way to protect themselves is to acquire nuclear weapons, while, in reality, it makes the world a more dangerous place.

The doctrine of MAD requires adversaries to have a credible nuclear capability. During the Cold War, this thinking spurred the Americans and the Soviets to build massive nuclear arsenals. However, as we discussed previously, even nations with limited nuclear capabilities can initiate a nuclear winter with an exchange of about a hundred World War II–size nuclear weapons. In addition to the millions who would perish in the nuclear exchange, humanity would face the prospect of worldwide hunger as the resultant layer of smoke and dust in the atmosphere would prevent the sun's rays from reaching the earth's surface.

Proponents of MAD argue it prevented nuclear war. They point to the Cold War era, when the United States and the Soviet Union each had over fourteen thousand missiles targeting each other's cities, leadership, and military capabilities.

Opponents use the same data to argue it was a delicate balance of terror, and each side almost launched its weapons on numerous occasions. In the book's introduction, I provide an example of one such incident. To briefly summarize, on September 26, 1983, Lt. Col. Stanislav Petrov, a Soviet ballistics officer commanding a bunker outside Moscow, chose to ignore what his early warning satellite system was telling him: the United States had launched five intercontinental ballistic missiles at his country. Had Petrov followed protocol and sounded the alarm of an impending attack, the incident could have escalated into a full-scale nuclear exchange.[4] However, Petrov used his judgment and asked himself why the United States would only launch five missiles if it was initiating a first strike. It did not make sense. Later, the Soviets learned that the sensors had misinterpreted the sun's reflection off cloud tops as missiles.

The United States has also had similar incidents. For example, on June 3, 1980, the North American Aerospace Defense Command (NORAD) center displayed large numbers of incoming missiles. NORAD prepared for retaliation but canceled it because NORAD officers judged that the numbers reported were too abnormal to be a missile attack. Later, NORAD technicians determined a faulty computer chip had produced random data that resulted in the deceptive displays.

Both the Soviets and the Americans experienced more near misses during the Cold War, but human wisdom prevented Armageddon. We need to pause and fully appreciate that avoiding a nuclear holocaust was the result of humans exercising their judgment. It had nothing to do with the sophistication of the weapons systems, the capabilities of computers, or the number of nuclear weapons in respective arsenals.

Now, let us fast forward to the present. What do we find?

The emergence of directed-energy weapons.

The development of autonomous weapons systems.

Let us explore how these technological capabilities are undermining the doctrine of MAD.

I can envision a time—and I believe that time may be years, not decades, away—when the United States will have laser weapons that can destroy an enemy's intercontinental ballistic missiles in flight. How could this be possible?

The trajectory of a nuclear-tipped ICBM has three phases:

1. The powered flight stage to boost the nuclear payload into outer space, typically about 1,200 miles above the earth

2. The free-flight part in outer space, where the payload orbits the earth until it reaches its target reentry point

3. The reentry portion, where the payload reenters the earth's atmosphere at supersonic speeds, typically about twenty times faster than the speed of sound[5]

At this point, the nuclear payload can explode above or at ground level, depending on its settings.

The ICBM moves the slowest during phase 1, which makes it an easier target for antiballistic missile weapons, such as missiles and space-based lasers. During phase 2, the payload is similar to an orbiting satellite. As previously discussed, the U.S. military demonstrated lasers could successfully destroy a low-Earth orbit satellite.[6] As the power of lasers increases and the targeting technology matures, lasers should be able to destroy the orbiting nuclear payloads.

This juncture is a good time to review briefly the Anti-Ballistic Missile Treaty between the United States and the Soviet Union that placed limits on the ABM systems used in defending areas against nuclear-tipped ballistic missiles. The agreement required each party to deploy not more than two antiballistic missile complexes, each limited to a hundred antiballistic missiles.[7] Signed in 1972, the treaty remained in force for the next thirty years, even after the Soviet Union's collapse in 1991.[8] A memorandum of understanding prepared in 1997 established Belarus, Kazakhstan, the Russian Federation, and Ukraine as the USSR's successor in the treaty.[9] However, as noted in chapter 1, President George W. Bush withdrew the United States from the treaty in 2002, leading to its termination and the formation of the U.S. Missile Defense Agency.[10]

Those who supported the withdrawal argue it enabled the United States to build a national missile defense to protect the country from nuclear blackmail by rogue states, such as Iran and North Korea, while critics contend it dealt a fatal blow to the Non-Proliferation Treaty, which is an international treaty to prevent the spread of nuclear weapons. Meanwhile, some potential adversaries feared the construction of a missile defense system would enable the United States to attack with a nuclear first strike. This thinking is what likely led Russian president Putin to respond to the U.S. withdrawal by ramping up Russia's nuclear capabilities.[11]

I believe the United States' primary reason to withdraw from the ABM Treaty was, in fact, out of concern that a rogue state might use nuclear weapons for political leverage or as a last-ditch strike during a conflict with the United States. Now, almost two decades later, I think the reason the United States is reluctant to support any ABM treaty is profoundly different—specifically, the belief that AI-directed, high-power lasers will provide a shield against all types of missiles and missile swarms. For example, the U.S. military already has proven it can destroy low-Earth orbit satellites using a laser weapon.[12] As noted in chapter 4, the United States has ordered two new lasers from Lockheed Martin that some sources estimate will be in the 150-kilowatt range. If those lasers have that type of power, I think they could destroy orbiting nuclear payloads, which are typically in a higher Earth orbit (about twelve hundred miles above Earth). Having only two such lasers may not immediately nullify the doctrine of MAD, but given time, the U.S. military is likely to have a laser on every navy destroyer. Additionally, according to the *Daily Beast*, "Under pressure from Congress and the Trump administration to bolster its military presence in space, the U.S. Defense Department is once again studying ways of mounting missile or lasers onto satellites in order to shoot down ballistic missiles before they can strike U.S. soil."[13]

Today, the United States leads in laser weapons and will likely be the first nation to deploy them widely. However, you may wonder, How would they defeat nuclear weapons delivered by hypersonic glide missiles? To a laser beam traveling at the speed of light, even a hypersonic glide vehicle appears to be standing still; it only requires radar to identify the missile's coordinates, and not its trajectory, to destroy the missile.

As we have discussed, this scenario is not science fiction. It is a matter of manufacturing lasers and incorporating them in U.S. military combat tactics. Based on all publicly available data, I think by the year 2030, the U.S. military will likely have enough laser capability to destroy all ICBMs and hypersonic glide vehicles in flight before they can obliterate U.S. cities and military assets.

If true, this will nullify the doctrine of MAD. However, I think Americans will still be reluctant to use nuclear weapons. By their nature, these weapons are indiscriminate, and the aftereffects of heat and radiation would cause unnecessary suffering. As a result, the 1977 Geneva Convention Protocols I and II—amendments to the Geneva Convention relating to the protection of victims of international and regional armed conflicts—consider them illegal weapons in warfare. The United States has signed but not ratified these protocols, so it is not bound by them.

I also think close-pacing potential adversaries such as China and Russia may acquire laser technology with similar capabilities in less than a decade after the U.S. military deploys them. That would be consistent with historical precedents. For example, the Soviet Union detonated its first atom bombs just four years after the U.S. military bombed Japan.

Denuclearization

When world militaries possess lasers that can destroy any nation's ability to deliver nuclear weapons, then serious dialogue would be possible on eliminating them. In a sense, nuclear weapons may go the way of World War II battleships, which the U.S. Navy phased out because they became obsolete in the face of jet aircraft and missile-carrying destroyers. This same fate is what I believe could happen to nuclear weapons: they will become outdated in the face of projected-energy weapons.

Another reason to eliminate nuclear weapons is the issue that arises with autonomous weapons in warfare. When the world came dangerously close to nuclear apocalypse on several occasions, only human judgment prevented the unthinkable from occurring. Now, as world militaries rely more heavily on autonomous weapons systems, computer algorithms are replacing human wisdom. To my eye, even when nations use weapons in a semiautonomous mode, c-war relegates human judgment to a secondary role, meaning the commander supervising the armament may not have the reaction time necessary to control it. Based on this understanding, even semiautonomous weapons could initiate an unintentional nuclear conflict before world leaders can react to stop it. In my opinion, autonomous weapons, even if used in a semiautonomous mode, make the world a more dangerous place.

Humans have shown the capacity to rid the world of weapons that have the potential to destroy humanity. One example is the Biological Weapons Convention. It requires the 183 states that are a party to it as of August 2019 to prohibit the development, production, and stockpiling of biological agents and toxins and their associated weapons.[14]

Given this historical precedent, I believe that humanity will abolish the use of nuclear weapons as new technology makes them useless, such as lasers rendering their delivery impossible, and the potential for their unintended use, due to autonomy, becomes a significant probability. Additionally, states will recognize that the detonation of nuclear weapons not only threatens the target but also more widely all of humanity with a nuclear winter and radiation fallout. Currently, the best organization to bring denuclearization to fruition is Global Zero, an international nonpartisan group of three hundred world leaders dedicated to achieving nuclear disarmament.[15]

Autonomous Directed-Energy Weapons

As the pace of war quickens, via the integration of computer technology in weapons systems, and as states deploy autonomous directed-energy weapons, the potential for nations to engage in c-war increases. Let us understand why.

As noted in chapters 3 and 4, both the Third and Fourth U.S. Offset Strategies place emphasis on integrating artificial intelligence into weapons systems. The U.S. military, along with other world militaries, views this strategy as the fastest way to "achieve a step increase in performance" of conventional weapons systems.[16] This thinking is ultimately leading nations to develop autonomous weapons, such as the U.S. Navy's x-47b, which is an autonomous jet-powered stealth drone intended for aircraft carrier-based operations (see chapter 10).

Autonomous weapons, by their nature, replace human judgment with computer algorithms. Even in a semiautonomous mode, the nature of c-war may relegate the commander supervising an autonomous weapon to the role of a spectator. Conceivably, two capable adversaries could engage in an unintentional c-war based solely on the actions of their autonomous weapons and without the governments of either nation making a conscious decision to engage in combat.

Consistent with current military strategy, nations will seek to automate their most effective armaments. To the Russians, this means using autonomous weapons with a nuclear warhead; however, the U.S. military plans to use autonomous weapons with a conventional warhead (see chapter 10). In the 2020s, I think lasers and microwave generators will be widely deployed by the U.S. military, because the technology to install high-power lasers and microwave generators will be widely available and highly effective in both offensive and defensive roles in that period.

Autonomous weapons in the 2020s will have an intelligent agent at their core, meaning they will be effective weapons but will not possess human judgment. They will follow their computer algorithms. Just as it is not possible to write a military policy that removes the need for human interpretation, it is currently not possible to write a computer algorithm that eliminates the need for human judgment. Therefore, I believe autonomous directed-energy weapons will make the world more dangerous. I also think that Russia's intent to deploy autonomous nuclear weapons is reckless and endangers humanity.

I do not believe there is a perfect solution to this issue; however, world militaries can take two steps toward lowering the potential of starting a c-war:

1. Require autonomous weapons to operate only under human supervision, thus deploying them as semiautonomous weapons.

2. Require autonomous weapons to incorporate only conventional warheads.

While these conditions will not altogether remove the danger of nations engaging in unintentional c-war, they will reduce the probability and destruction of such a conflict. To my mind, they represent the minimum first steps to deploying autonomous weapons wisely. The good news is that U.S. military policy already mandates these requirements. The challenge is to get the Russians and the Chinese to do likewise.

Concluding Remarks

This chapter makes a crucial point: as the pace of war quickens and nations deploy autonomous directed-energy weapons, the potential for countries to engage in c-war increases. To prevent this possibility, I offer three guidelines:

1. Eliminate nuclear weapons. I believe this will be possible if directed-energy weapons, such as lasers, render their delivery impossible.

2. Use autonomous weapons only under human supervision, making them semiautonomous.

3. Arm autonomous weapons only with conventional warheads.

I recognize the implementation of these guidelines will be contentious and difficult. I offer them as a starting point for an informed discussion. I invite dialogue and challenge leaders at every level to engage in endeavors to prevent c-war.

The planet Earth is our home. While one day we may have the capability to colonize other planets, we are more likely to have the ability to engage in c-war long before that occurs. Therefore, we all have a stake in keeping Earth inhabitable. We need to recognize c-war has the potential to render Earth a barren wasteland. Thus, our current reality rests with the wisdom of our world leaders.

Before we conclude, let me leave you with these words of wisdom from Calvin Coolidge, who served as the thirtieth president of the United States from 1923 to 1929: "Knowledge comes, but wisdom lingers. It may not be difficult to store up in the mind a vast quantity of facts within a comparatively short time, but the ability to form judgments requires the severe discipline of hard work and the tempering heat of experience and maturity."[17]

Life is a journey, and I hope you enjoyed yours with me through this book. Even more than that, I want this book to be a starting point for informed thinking and action to avoid engaging in c-war.

I commend you for seeking knowledge on c-war. Now, I would like to ask you to take the next step. Share this book and its contents with others. I think the path to wisdom and maturity is for people everywhere to be educated on the unique dangers and opportunities our new technology has wrought. Armed with this knowledge, we share a common goal—to ensure future generations inherit a safer world.

I hope that someday our paths will cross. My deepest desire is that day finds us in a more peaceful world, where nations work together to further the ascent of humanity.

APPENDIX A

U.S. and Chinese Defense Budgets Adjusted for
Purchasing Power Parity and Labor Costs

Two factors not reflected in a nation's defense spending are its purchasing power parity and labor costs. We can understand these elements by comparing China's defense budget to that of the United States while using a snapshot of their military budgets in 2017. As complete data for later years becomes available to make similar comparisons, I believe the conclusions drawn here will remain unchanged.

In 2017 China's gross domestic product (GDP) was $12.0 trillion, and the United States' GDP was $19.8 trillion.[1] In examining their defense budgets relative to their GDPs, we first convert everything to U.S. dollars at market rates (i.e., the price to buy something in the world market with U.S. dollars):

China: $227.8 billion defense budget, 1.9 percent of GDP

United States: $606.0 billion defense budget, 3.1 percent[2]

At first glance, the United States appears to spend almost three times more on defense than China does. However, simply comparing this raw data is misleading. We need to review these budgets in terms of purchasing power parity and labor costs.

Purchasing Power Parity: This is a measure of how much a particular currency will buy in its regional market relative to that of another currency. Two currencies are equal when a basket of goods costs the same in both countries after the exchange rate. For example, if a basket of products, such as pencils and paper, costs $100 in the United

States but only $50 in China (after currency exchange), then each Chinese dollar is equal to two U.S. dollars in terms of purchasing power parity. What does that mean regarding the defense budgets of China versus the United States? Since China buys most of its military equipment from domestic suppliers, which are either officially government owned or heavily government influenced, its products are generally less expensive than their U.S. equivalents. According to *Breaking Defense*, a digital magazine on the strategy, politics, and technology of defense, China's defense budget in 2017 was $434.5 billion versus the U.S. defense budget of $606.0 billion after adjusting for purchasing power parity.[3] Now, we would conclude that the U.S. defense budget was about 50 percent higher than China's.

Labor Costs: This is the percentage of the defense budget allocated for pay and benefits. Again, according to *Breaking Defense*, 42 percent of the U.S. defense budget goes to pay and benefits.[4] Therefore, the U.S. budget, minus personnel costs, is $356 billion. Unfortunately, we do not have a good measure of what China spends on personnel; however, according to *Breaking Defense*, it is a relatively small share of its budget.[5] For comparison, let us assume China spends half (i.e., 21 percent) of what the U.S. military spends on labor. Then China's defense budget, less personnel, would be $343.2 billion. Thus, we can conclude that China is spending about the same on defense as the United States.

The bottom line is that both countries spend about the same amount on national security. From this analysis, we can draw two implications:

1. If China's economy continues to grow, it will be able to outspend the U.S. military on defense and potentially develop better weapons.

2. Due to its population, which is about three times that of the United States, China will be able to field a more massive military.

However, the United States still enjoys a lead in military technology and weapon sophistication. Thus, every dollar the U.S. government spends, builds on this lead. Every dollar that China spends, meanwhile, works to close the technology and weapons gap.

APPENDIX B

The Design and Operation of a Laser

The word "laser" is an acronym for light amplification by stimulated emission of radiation. Now, it is only natural to ask what that means. I address this question in two parts—by explaining the design of a laser and by demonstrating how a laser works to form a laser beam.

Conceptually, laser design is relatively simple and only consists of three components:

1. A laser medium. This is the source of light within a laser. It can be composed of a solid, gas, or liquid. The medium is typically confined in a tube.

2. An energy source. The energy source can be an electrical current or another laser.

3. A feedback element. This component takes the output from the laser medium and routes it back into the system. (The reason for this will be apparent when we examine how the laser works.) Most often, the feedback element consists of two mirrors that confine the laser light within the laser medium. In practice, one mirror is partially silvered, meaning the silver coating is less dense and can allow the most powerful laser light to escape.[1]

These components work together to form a laser beam.

Forming a laser beam occurs in three steps:[2]

1. The laser medium absorbs energy from the energy source and becomes excited; that is, the electrons associated with the atoms

that form the medium move from a lower-energy orbit to a higher one. (*Note:* You can think of an atom as a small solar system. In the place of the sun is a nucleus, typically consisting of the sub-atomic particles neutrons and protons. Instead of planets, sub-atomic particles called electrons orbit the nucleus. The electrons close to the nucleus, or in a low orbit, have lower energy than those farther away.)

2. The higher orbit assumed by the electrons is unstable, so they seek to return to their "ground" state, or their original orbit. As the electrons transition to their ground state, they emit their absorbed energy in the form of photons, or light waves. All the light waves produced have the same wavelength and are "coherent," meaning they are identical.

3. The feedback element typically comprises a mirror at each end of the laser medium, with one being partially silvered. The mirrors trap the emitted coherent light, causing it to reflect back and forth in the tube containing the medium. This back-and-forth motion of coherent light stimulates more electrons in the system and, in turn, produces more light of the same wavelength. This process amplifies the initial low-energy pulse quadrillions of times, resulting in highly energetic coherent light. The partly silvered mirror allows some of the coherent light to escape and form a highly energetic laser beam.

APPENDIX C

Radiation-Hardened Electronics and System Shielding Resources

Radiation-Hardened Electronics

Honeywell Aerospace. "Radiation Hardened SOI CMOS Technology." Accessed July 28, 2019. https://aerospace.honeywell.com
/content/dam/aero/en-us/documents/learn/products/sensors
/brochures/RadiationHardenedSOICMOSTechnology-bro.pdf.

Keller, John. "The Evolving World of Radiation-Hardened Electronics." *Military & Aerospace Electronics*, June 1, 2018. https://
www.militaryaerospace.com/computers/article/16707204/the
-evolving-world-of-radiationhardened-electronics.

Verbeeck, Jens. "Protecting Electronics from the Effects of Radiation." *Nuclear Engineering International*, August 3, 2017. https://
www.neimagazine.com/features/featureprotecting-electronics
-from-the-effects-of-radiation-5890599.

Radiation System Shielding

BCC Research. "Radiation Shielding and Monitoring: Technologies
and Application Markets, 2019 Report." Research and Markets,
May 28, 2019. https://www.researchandmarkets.com/reports
/4771277/radiation-shielding-and-monitoring-technologies
#pos-0. This comprehensive report requires purchase.

McKay, Chris, and Bin Chen. "Radiation Shielding System Using
a Composite of Carbon Nanotubes Loaded with Electropolymers." Tech Briefs, March 1, 2012. https://www.techbriefs.com
/component/content/article/tb/techbriefs/materials/13046.

NASA. "Technology Opportunity: Radiation Shielding Systems Using Nanotechnology: A System for Shielding Personnel and/or Equipment from Radiation Particles." August 8, 2011. https://www.nasa.gov/pdf/579158main_radiation%20shielding%20systems.pdf.

APPENDIX D

Articles Describing the Operation of Nonnuclear EMP Devices

Encyclopaedia Britannica. S.v., "pinch effect." Accessed May 19, 2020. https://www.britannica.com/science/pinch-effect. Explains the pinch effect, which is fundamental in creating electromagnetic field devices.

Harris, Tom. "How E-Bombs Work." How Stuff Works, March 23, 2003. https://science.howstuffworks.com/e-bomb3.htm. Provides a simple explanation of flux compression generators.

IDC Technologies. "Marx Generator." Accessed May 19, 2020. http://www.idc-online.com/technical_references/pdfs/electronic_engineering/Marx_Generator.pdf. Explains the operation of a Marx generator.

Younger, Stephen, Irvin Lindemuth, Robert Reinovsky, C. Maxwell Fowler, James Goforth, and Carl Ekdahl. "Scientific Collaborations between Los Alamos and Arzamas-16 Using Explosive-Driven Flux Compression Generators." *Los Alamos Science* 24 (1996): 48–71. https://fas.org/sgp/othergov/doe/lanl/pubs/00326620.pdf. Describes the operation of Andrei Sakharov's 1951 concept of flux compression generators.

NOTES

Introduction

1. Sydney J. Freedberg Jr., "Hagel Lists Key Technologies for US Military; Launches 'Offset Strategy,'" *Breaking Defense*, November 16, 2014, https://breakingdefense.com/2014/11/hagel-launches-offset-strategy-lists-key-technologies.

2. Cheryl Pellerin, "Deputy Secretary: Third Offset Strategy Bolsters America's Military Deterrence," DOD *News*, October 31, 2016, https://www.defense.gov/Explore/News/Article/Article/991434/deputy-secretary-third-offset-strategy-bolsters-americas-military-deterrence.

3. Sun Tzu, *The Art of War*, https://www.goodreads.com/quotes/102241-if-your-enemy-is-secure-at-all-points-be-prepared (accessed May 6, 2020).

4. "DF-26 (Dong Feng-26)," *Missile Threat*, January 31, 2020, https://missilethreat.csis.org/missile/dong-feng-26-df-26. See also Hans M. Kristensen, "China's New DF-26 Missile Shows Up at Base in Eastern China," Federation of American Scientists, January 21, 2020, https://fas.org/blogs/security/2020/01/df-26deployment/.

5. "Turkey and Russia Cosy Up over Missiles," *Economist*, May 4, 2017, https://www.economist.com/europe/2017/05/04/turkey-and-russia-cosy-up-over-missiles.

6. Darrell M. West and John R. Allen, "How Artificial Intelligence Is Transforming the World," Brookings Institution, April 24, 2018, https://www.brookings.edu/research/how-artificial-intelligence-is-transforming-the-world.

7. Doug Irving, "How Artificial Intelligence Could Increase the Risk of Nuclear War," RAND Corporation, April 23, 2018, https://www.rand.org/blog/articles/2018/04/how-artificial-intelligence-could-increase-the-risk.html.

1. Tilting the Balance of Terror

1. Mark Parillo, *Why Air Forces Fail: The Anatomy of Defeat* (Lexington: University Press of Kentucky, 2006), 288.

2. Mark Chambers, *Wings of the Rising Sun: Uncovering the Secrets of Japanese Fighters and Bombers of World War II* (Oxford: Osprey Publishing, 2018), 282.

3. "The Japanese Attacked Pearl Harbor, December 7, 1941," Library of Congress, America's Story, http://www.americaslibrary.gov/jb/wwii/jb_wwii_pearlhar_1.html (accessed December 17, 2018).

4. David Martin, "The Pentagon's Ray Gun," *60 Minutes*, CBS, February 29, 2008, https://www.cbsnews.com/news/the-pentagons-ray-gun.

5. Allison Barrie, "'Force Field' Technology Could Make U.S. Tanks Unstoppable," Fox News, August 2, 2018, https://www.foxnews.com/tech/force-field-technology-could-make-us-tanks-unstoppable.

6. David Allen Batchelor, "The Science of Star Trek," NASA, July 20, 2016, https://www.nasa.gov/topics/technology/features/star_trek.html.

7. "America's Utilities Prepare for a Nuclear Threat to the Grid," *Economist*, September 9, 2017, https://www.economist.com/business/2017/09/09/americas-utilities-prepare-for-a-nuclear-threat-to-the-grid.

8. Habib Siddiqui, "Letter from America—World Disorder: Reform at the UN Needed," *Asian Tribune*, October 28, 2018, http://asiantribune.com/node/92261.

9. Anne Barnard and Karam Shoumali, "U.S. Weaponry Is Turning Syria into Proxy War with Russia," *New York Times*, October 12, 2015, https://www.nytimes.com/2015/10/13/world/middleeast/syria-russia-airstrikes.html?_r=0; and Mark Landler, Helene Cooper, and Eric Schmitt, "Trump Withdraws U.S. Forces from Syria, Declaring 'We Have Won against ISIS,'" *New York Times*, December 19, 2018, https://www.nytimes.com/2018/12/19/us/politics/trump-syria-turkey-troop-withdrawal.html.

10. Henry D. Sokolski, ed., *Getting Mad: Nuclear Mutual Assured Destruction, Its Origins and Practice* (Carlisle PA: Strategic Studies Institute Home, November 2004), https://www.globalsecurity.org/jhtml/jframe.html#https://www.globalsecurity.org/wmd/library/report/2004/ssi_sokolski.pdf|||Getting%20MAD:%20Nuclear%20Mutual%20Assured%20Destruction,%20Its%20Origins%20and%20Practice.

11. Keir A. Lieber and Daryl G. Press, "The End of MAD? The Nuclear Dimension of U.S. Primacy," *International Security* 30, no. 4 (Spring 2006): 7–8, https://www.mitpressjournals.org/doi/pdf/10.1162/isec.2006.30.4.7.

12. Alexander Velez-Green, "We Need Mutually Assured Destruction," *National Review*, May 1, 2017, https://www.nationalreview.com/2017/05/mutually-assured-destruction-obsolescent-and-could-be-very-bad.

13. Sebastien Roblin, "This Old U.S. Navy Nuclear Submarine Could Nuke 24 Cities (in One Shot)," *National Interest*, May 13, 2019, https://nationalinterest.org/blog/buzz/old-us-navy-nuclear-submarine-could-nuke-24-cities-one-shot-57332.

14. Bill Gertz, "Pentagon: China, Russia Soon Capable of Destroying U.S. Satellites," *Washington Free Beacon*, January 30, 2018, https://freebeacon.com/national-security/pentagon-china-russia-soon-capable-destroying-u-s-satellites.

15. "Fact Sheet: Ballistic vs. Cruise Missiles," Center for Arms Control and Non-Proliferation, April 27, 2017, https://armscontrolcenter.org/fact-sheet-ballistic-vs-cruise-missiles.

16. "Treaty on Principles Governing the Activities of States in the Exploration and Use of Outer Space, Including the Moon and Other Celestial Bodies," U.S. Department of State, Bureau of Arms Control, Verification, and Compliance, October 10, 1967, https://www.state.gov/t/isn/5181.htm.

17. Jeff Foust, "U.S. Dismisses Space Weapons Treaty Proposal as 'Fundamentally Flawed,'" *SpaceNews*, September 11, 2014, https://spacenews.com/41842us-dismisses-space-weapons-treaty-proposal-as-fundamentally-flawed.

18. Velez-Green, "We Need."

19. Dictionary.com, s.v. "rail gun," https://www.dictionary.com/browse/rail-gun (accessed October 28, 2018); and David Martin, "Navy's Newest Weapon Kills at Seven Times the Speed of Sound," CBS *Evening News*, April 7, 2014, https://www.cbsnews.com/news/navys-newest-weapon-kills-at-seven-times-the-speed-of-sound.

20. Shawn Snow, "U.S. May Field Railgun on Zumwalt Destroyer," *Diplomat*, March 1, 2016, https://thediplomat.com/2016/03/u-s-may-field-railgun-on-zumwalt-destroyer.

21. Snow, "U.S. May Field."

22. Martin, "Navy's Newest Weapon."

23. Velez-Green, "We Need."

24. Amanda Macias, "China Just Tested the World's Most Powerful Naval Gun, and US Intelligence Says It Will Be Ready for Warfare by 2025," CNBC, January 30, 2019, https://www.cnbc.com/2019/01/30/china-naval-gun-ready-for-warfare-by-2025-us-intelligence.html.

25. Matt Stroud, "Inside the Race for Hypersonic Weapons," *Verge*, March 6, 2018, https://www.theverge.com/2018/3/6/17081590/hypersonic-missiles-long-range-arms-race-putin-speech.

26. "Russia Deploys Avangard Hypersonic Missile System," BBC News, December 27, 2019, https://www.bbc.com/news/world-europe-50927648.

27. Stroud, "Inside the Race."

28. Stroud, "Inside the Race."

29. Stroud, "Inside the Race."

30. Velez-Green, "We Need."

31. Franz-Stefan Gady, "World's Largest Anti-Submarine Robot Ship Joins U.S. Navy," *Diplomat*, February 6, 2018, https://thediplomat.com/2018/02/worlds-largest-anti-submarine-robot-ship-joins-us-navy.

32. "ACTUV 'Sea Hunter' Prototype Transitions to Office of Naval Research for Further Development," Defense Advanced Research Projects, Agency News and Events, January 30, 2018, https://www.darpa.mil/news-events/2018-01-30a.

33. Velez-Green, "We Need."

2. The Quest for Global Dominance

1. "Closer Than Ever: It Is 100 Seconds to Midnight," *Bulletin of the Atomic Scientists*, January 23, 2020, https://thebulletin.org/doomsday-clock/.

2. Tom Nichols, "Five Ways a Nuclear War Could Still Happen," *National Interest*, June 16, 2014, https://nationalinterest.org/feature/five-ways-nuclear-war-could-still-happen-10665?page=0%2C1.

3. Robert Monroe, "Only Trump Can Restore America's Ability to Win a Nuclear War," *Hill*, January 26, 2018, https://thehill.com/opinion/national-security/370901-only-trump-can-restore-americas-ability-to-win-a-nuclear-war.

4. Harry Atkins, "How Many People Died in the Hiroshima and Nagasaki Bombings?," *HistoryHit*, August 9, 2018, https://www.historyhit.com/how-many-people-died-in-the-hiroshima-and-nagasaki-bombings.

5. "Tons (Explosives) to Gigajoules Conversion Calculator," UnitConversion.org, http://www.unitconversion.org/energy/tons-explosives-to-gigajoules-conversion.html (accessed February 3, 2019).

6. "The B83 (Mk-83) Bomb," Nuclear Weapon Archive, http://nuclearweaponarchive.org/Usa/Weapons/B83.html (accessed February 3, 2019).

7. "The Atomic Bombings of Hiroshima and Nagasaki: Summary of Damages and Injuries," Atomic Archive, http://www.atomicarchive.com/Docs/MED/med_chp3.shtml (accessed June 30, 2019).

8. Michelle Hall, "By the Numbers: World War II's Atomic Bombs," CNN Library, August 6, 2013, https://www.cnn.com/2013/08/06/world/asia/btn-atomic-bombs/index.html; and "Atomic Bombings."

9. Fiona MacDonald, "Watch: How Far Away Would You Need to Be to Survive a Nuclear Blast?," Science Alert, January 30, 2017, https://www.sciencealert.com/watch-how-far-away-would-you-need-to-be-to-survive-a-nuclear-blast.

10. Amanda Macias, "There Are about 14,500 Nuclear Weapons in the World; Here Are the Countries That Have Them," CNBC, March 16, 2018, https://www.cnbc.com/2018/03/16/list-of-countries-with-nuclear-weapons.html.

11. U.S. Congress, Office of Technology Assessment, "The Effects of Nuclear War" (Washington DC: U.S. Government Printing Office, May 1979), https://ota.fas.org/reports/7906.pdf.

12. United States Census Bureau, https://www.census.gov/en.html (accessed June 24, 2019).

13. "Demographics of Russia," Wikipedia, https://en.wikipedia.org/wiki/Demographics_of_Russia (accessed June 24, 2019).

14. Peter Vincent Pry, "Nuclear EMP Attack Scenarios and Combined-Arms Cyber Warfare: Report to the Commission to Assess the Threat to the United States from Electromagnetic Pulse (EMP) Attack," July 2017, http://www.firstempcommission.org/uploads/1/1/9/5/119571849/nuclear_emp_attack_scenarios_and_combined-arms_cyber_warfare_by_peter_pry_july_2017.pdf.

15. Pry, "Nuclear EMP Attack."

16. Alan Robock and Owen Brian Toon, "Self-assured Destruction: The Climate Impacts of Nuclear War," *Bulletin of the Atomic Scientists*, September 1, 2012, https://thebulletin.org/2012/09/self-assured-destruction-the-climate-impacts-of-nuclear-war.

17. Alan Robock, Luke Oman, and Georgiy L. Stenchikov, "Nuclear Winter Revisited with a Modern Climate Model and Current Nuclear Arsenals: Still Catastrophic Consequences," *Journal of Geophysical Research* 112 (April 27, 2007), https://doi.org/10.1029/2006jd008235.

18. M. Harwell and C. Harwell, "Nuclear Famine: The Indirect Effects of Nuclear War," in *The Medical Implications of Nuclear War*, ed. F. Solomon and R. Marston (Washington DC: National Academy Press, 1986), 117–35.

19. Michael J. Mills et al., "Multidecadal Global Cooling and Unprecedented Ozone Loss Following a Regional Nuclear Conflict," *AGU Publications*, February 7, 2014, https://doi.org/10.1002/2013EF000205.

20. Mills et al., "Multidecadal Global Cooling."

21. Wayne M. Dzwonchyk and John Ray Skates, "A Brief History of the U.S. Army in World War II" (Washington DC: U.S. Army Center of Military History, 1992), https://history.army.mil/brochures/brief/overview.htm#3.

22. Jim Garamone, "Dunford: Speed of Military Decision-Making Must Exceed Speed of War," *DOD News*, January 31, 2017, https://www.defense.gov/Explore/News/Article/Article/1066045/dunford-speed-of-military-decision-making-must-exceed-speed-of-war.

23. Mike Lehr, "Top Seven Sun Tzu Quotes: #5 Speed in Decision Making," Omega Z Advisors, February 8, 2019, https://omegazadvisors.com/2013/05/27/top-seven-sun-tzu-quotes-5-speed-in-decision-making/.

24. David Lague, "China Leads U.S. on Potent Super-fast Missiles," Reuters, April 25, 2019, https://www.reuters.com/article/us-china-army-hypersonic/china-leads-u-s-on-potent-super-fast-missiles-idUSKCN1S11E6.

25. Amanda Macias, "Putin Says Russia Will Deploy Hypersonic Missiles in 'Coming Months,' Surpassing U.S. and China," *CNBC*, October 18, 2018, https://www.cnbc.com/2018/10/18/putin-says-hypersonic-missiles-will-deploy-in-coming-months.html.

26. Iain Boyd, "U.S., Russia, China Race to Develop Hypersonic Weapons," Phys.org, May 1, 2019, https://phys.org/news/2019-05-russia-china-hypersonic-weapons.html.

27. Scott McDonald, "U.S. Navy to Develop Laser Weapon That Can 'Destroy' Small Boats, Hostile Drones," *Newsweek*, March 20, 2019, https://www.newsweek.com/navy-develop-laser-weapon-can-destroy-small-boats-hostile-drones-1370648.

28. Steven Aftergood, "WWII Atomic Bomb Project Had More Than 1,500 'Leaks,'" Federation of American Scientists' *Secrecy News*, August 21, 2014, https://fas.org/blogs/secrecy/2014/08/manhattan-project-leaks.

29. U.S. Congress, House of Representatives, Committee on Military Affairs, "Atomic Energy," Hearings on H.R. 4280, "An Act for the Development and Control of Atomic Energy," October 9 and 18, 1945, 79th Cong., 1st Sess., 1945, Testimony of Gen. L. R. Groves at p. 20, quoted in Arvin Quist, *Principles for Classification of Information*, vol. 2 of *Security Classification of Information* (Oak Ridge TN: Oak Ridge National Laboratory, April 1993), chap. 8, https://fas.org/sgp/library/quist2/chap_8.html.

30. "Transcript of President Dwight D. Eisenhower's Farewell Address (1961)," https://www.ourdocuments.gov/doc.php?flash=false&doc=90&page=transcript (accessed January 25, 2019).

31. "Military Expenditure (% of GDP): All Countries and Economies," World Bank, 2018, https://data.worldbank.org/indicator/ms.mil.xpnd.gd.zs.

32. "NIF Sets New Laser Energy Record," Lawrence Livermore National Laboratory, July 10, 2018, https://www.llnl.gov/news/nif-sets-new-laser-energy-record.

33. "FAQs," Lawrence Livermore National Laboratory, National Ignition Facility & Photon Science, https://lasers.llnl.gov/about/faqs (accessed February 27, 2019).

34. "FAQs."

35. Boyd, "U.S., Russia, China."

36. Robert Bruce, "The U.S. Navy's Electric Weaponry," *Small Arms Defense Journal*, February 19, 2016, http://www.sadefensejournal.com/wp/the-us-navys-electric-weaponry/.

3. The Fourth U.S. Offset Strategy

1. "Take a Closer Look: America Goes to War," National World War II Museum, https://www.nationalww2museum.org/students-teachers/student-resources/research-starters/america-goes-war-take-closer-look (accessed February 28, 2019).

2. W. Gardner Selby, "U.S. Army Was Smaller Than the Army for Portugal before World War II," *Politifact*, June 13, 2014, https://www.politifact.com/texas/statements/2014/jun/13/ken-paxton/us-army-was-smaller-army-portugal-world-war-ii.

3. "Take a Closer Look."

4. *Encyclopaedia Britannica*, s.v. "Kim Il-Sung," https://www.britannica.com/biography/Kim-Il-Sung (accessed March 3, 2019); and *Encyclopaedia Britannica*, s.v. "Syngman Rhee," https://www.britannica.com/biography/Syngman-Rhee (accessed March 3, 2019).

5. Robert A. Divine et al., *Since 1865*, vol. 2 of *America Past and Present*, 8th ed. (New York: Pearson Longman, 2007), 819–21.

6. "U.N. Approves Armed Force to Repel North Korea," *Daily Sun*, June 27, 2015, https://www.daily-sun.com/arcprint/details/54006/2015/06/27/U.N.-approves-armed-force-to-repel-North-Korea/2015-06-27.

7. Peter Grier, "The First Offset," *Air Force Magazine*, June 2016, http://www.airforcemag.com/MagazineArchive/Magazine%20Documents/2016/June%202016/0616offset.pdf.

8. Grier, "The First Offset."

9. Robert S. Norris and Hans M. Kristensen, "Global Nuclear Weapons Inventories, 1945–2010," *Bulletin of the Atomic Scientists*, November 2, 2016, https://journals.sagepub.com/doi/abs/10.2968/066004008.

10. Robert Tomes, "The Cold War Offset Strategy: Assault Breaker and the Beginning of the RSTA Revolution," *War on the Rocks*, November 20, 2014, https://warontherocks.com/2014/11/the-cold-war-offset-strategy-assault-breaker-and-the-beginning-of-the-rsta-revolution.

11. Zachary Keck, "A Tale of Two Offset Strategies," *Diplomat*, November 18, 2014, https://thediplomat.com/2014/11/a-tale-of-two-offset-strategies.

12. Robert R. Tomes, "U.S. Defence Strategy from Vietnam to Operation Iraqi Freedom: Military Innovation and the New American War of War, 1973–2003" (Abingdon: Routledge, 2006), https://books.google.com/books/about/U.S._Defence_Strategy_from_Vietnam_to_Oper.html?id=7fySAgAAQBAJ.

13. Paul McLeary, "The Pentagon's Third Offset May Be Dead, but No One Knows What Comes Next," *Foreign Policy*, December 18, 2017, https://foreignpolicy.com/2017/12/18/the-pentagons-third-offset-may-be-dead-but-no-one-knows-what-comes-next.

14. Defense Secretary Ash Carter, "Remarks Previewing the FY 2017 Defense Budget," remarks at the Economic Club, Washington DC, February 2, 2016, https://dod.defense.gov/News/Speeches/Speech-View/Article/648466/remarks-previewing-the-fy-2017-defense-budget.

15. Aaron Mehta, "Strategic Capabilities Office Preparing for New Programs, Next Administration," *Defense News*, September 9, 2016, https://www.defensenews.com/pentagon/2016/09/09/strategic-capabilities-office-preparing-for-new-programs-next-administration.

16. Pellerin, "Deputy Secretary."

17. Caitlin Thorn, "Drop Zone: The Third Offset and Implications for the Future Operating Environment," *Over the Horizon*, January 19, 2018, https://othjournal.com/2018/01/19/drop-zone-the-third-offset-and-implications-for-the-future-operating-environment.

18. Jackie Northam, "China Makes a Big Play in Silicon Valley," *Weekend Edition Sunday*, NPR, October 7, 2018, https://www.npr.org/2018/10/07/654339389/china-makes-a-big-play-in-silicon-valley.

19. Michael Miklaucic, "America's Allies: The Fourth Strategic Offset," *Hill*, October 24, 2018, https://thehill.com/opinion/national-security/412656-americas-allies-the-fourth-strategic-offset.

20. Miklaucic, "America's Allies."

21. Jim Mattis, "Summary of the 2018 National Defense Strategy of the United States of America" (Washington DC: Department of Defense, 2019), 1, https://dod.defense.gov/Portals/1/Documents/pubs/2018-National-Defense-Strategy-Summary.pdf.

22. Mattis, "Summary," 3.

23. Mattis, "Summary," 8.

4. Laser Weapons

1. H. G. Wells, *War of the Worlds* (London: William Heinemann, 1899), 45, https://books.google.com/books?id=bQlHAAAAYAAJ&q=It+is+still+a+matter+of+wonder+how+the+Martians+are+able+to+slay+men+so+swiftly+and+so+silently.+#v=snippet&q=It%20is%20still%20a%20matter%20of%20wonder%20how%20the%20Martians%20are%20able%20to%20slay%20men%20so%20swiftly%20and%20so%20silently.&f=false.

2. Ernie Tretkoff, "Einstein Predicts Stimulated Emission," *American Physical Society News* 14, no. 8 (August/September 2005), https://www.aps.org/publications/apsnews/200508/history.cfm.

3. "Maser," Wikipedia, https://en.wikipedia.org/wiki/Maser#cite_note-autogenerated1-1 (accessed June 27, 2019); and Charles H. Townes, "Production of Coherent Radiation by Atoms and Molecules," Nobel Lecture, December 11, 1964, https://www.nobelprize.org/prizes/physics/1964/townes/lecture.

4. "Maser."

5. Joan Bromberg, "Oral Interview with Joseph Weber," American Institute of Physics, Oral History Interviews, University of Maryland, April 8, 1983, https://www.aip.org/history-programs/niels-bohr-library/oral-histories/4941.

6. J. P. Gordon, H. J. Zeiger, and C. H. Townes, "The Maser—New Type of Microwave Amplifier, Frequency Standard, and Spectrometer," *Physical Review* 99, no. 4 (August 1995): 1264–74, http://adsabs.harvard.edu/abs/1955PhRv...99.1264G.

7. "Nobel Prize in Physics 1964," Nobel Media AB 2020, https://www.nobelprize.org/prizes/physics/1964/summary/ (accessed April 13, 2020).

8. Mario Bertolotti, *Masers and Lasers: An Historical Approach*, 2nd ed. (Boca Raton: CRC Press, 2015), 89–91, https://books.google.com/books?id=4i_OBgAAQBAJ.

9. Mario Bertolotti, *History of the Laser* (Philadelphia: Institute of Physics Publishing, 1999), 181, https://books.google.com/books?id=JObDnEtzMJUC&pg=PA181&lpg=PA181 &dq=did+Charles+Townes+attend+the+seminar+given+by+weber+at+RCA&source= bl&ots=tzN2hy4aM7&sig=ACfU3U3FBX_tACKmKpyB-CI0FPds2x5BCQ&hl=en&sa =X&ved=2ahUKEwjGtZHRuKPjAhW1ds0KHSQODPwQ6AEwAHoECAoQAQ#v= onepage&q=did%20Charles%20Townes%20attend%20the%20seminar%20given%20by %20weber%20at%20RCA&f=false.

10. "Bright Idea: The First Laser," American Institute of Physics, https://history.aip.org /exhibits/laser/sections/whoinvented.html (accessed June 27, 2019).

11. "First Laser."

12. "First Laser."

13. "First Laser."

14. "First Laser."

15. "First Laser."

16. "First Laser."

17. "Case Study: Lasers," Institute of Physics, https://www.iop.org/cs/page_43644 .html (accessed June 29, 2019).

18. Dr. Rüdiger Paschotta, "Laser Applications," RP *Photonics Encyclopedia*, https:// www.rp-photonics.com/laser_applications.html (accessed June 29, 2019).

19. "Retinal Detachment Repair," Healthline, https://www.healthline.com/health/retinal -detachment-repair (accessed June 29, 2019); and "A Guide to Refractive and Laser Eye Surgery," WebMD, https://www.webmd.com/eye-health/overview-refractive-laser-eye -surgery#1 (accessed June 29, 2019).

20. Paschotta, "Laser Applications."

21. "Military Applications of Lasers," Science Clarified, http://www.scienceclarified .com/scitech/Lasers/Military-Applications-of-Lasers.html (accessed June 29, 2019).

22. "Military Applications."

23. "Military Applications."

24. "Military Applications."

25. Brian Resnick and *National Journal*, "A Brief History of Militarized Lasers," *Atlantic*, December 12, 2014, https://www.theatlantic.com/politics/archive/2014/12/a-brief -history-of-militarized-lasers/453453.

26. Michael Schrage, "Beam Can Blind Enemy Troops," *Washington Post*, December 17, 1983, https://www.washingtonpost.com/archive/politics/1983/12/17/beam-can-blind -enemy-troops/6d8fd0f4-3c92-4e4c-8d33-2ab1ab96d374/?utm_term=.167d9104f0ae.

27. "Blinding Laser Weapons: Questions and Answers," International Committee of the Red Cross, November 16, 1994, https://www.icrc.org/en/doc/resources/documents /misc/57jmcz.htm.

28. *Weapons Law Encyclopedia*, s.v. "1995 Protocol on Blinding Laser Weapons," http:// www.weaponslaw.org/instruments/1995-protocol-on-blinding-laser-weapons (accessed July 3, 2019).

29. United Nations, "Chapter XXVI, Disarmament," Treaty Collection, Vienna, October 13, 1995, https://treaties.un.org/pages/ViewDetails.aspx?src=Treaty&mtdsg_no=Xxvi-2-a&chapter=26&lang=en.

30. Schrage, "Beam Can Blind."

31. "Strategic Defense Initiative (SDI)," Atomic Heritage Foundation, July 18, 2018, https://www.atomicheritage.org/history/strategic-defense-initiative-sdi.

32. "Military Applications."

33. Andrew Glass, "Reagan Brands Soviet Union 'Evil Empire,'" *Politico*, March 8, 1983, https://www.politico.com/story/2018/03/08/this-day-in-politics-march-8-1983-440258.

34. Resnick and *National Journal*, "Brief History."

35. Associated Press, "U.S. Military Uses Laser 'Dazzler' to Stop Iraqis Who Ignore Warnings to Stop," NewsLibrary.com, May 18, 2006, http://nl.newsbank.com/nl-search/we/Archives?p_product=NewsLibrary&p_multi=APAB&d_place=APAB&p_theme=newslibrary2&p_action=search&p_maxdocs=200&p_topdoc=1&p_text_direct-0=11200C0C5BA16025&p_field_direct-0=document_id&p_perpage=10&p_sort=YMD_date:D&s_trackval=GooglePM.

36. "Over 6 Decades of Continued Transistor Shrinkage, Innovation," Intel Corporation press release, Santa Clara CA, May 1, 2011, https://www.intel.com/content/www/us/en/silicon-innovations/standards-22-nanometers-technology-backgrounder.html.

37. Ray Kurzweil, "The Law of Accelerating Returns," Kurzeweil Library, March 7, 2001, https://www.kurzweilai.net/the-law-of-accelerating-returns.

38. Mark Thompson, "*Zap Wars: U.S. Navy Successfully Tests Laser Weapon in the Persian Gulf*," *Time*, December 10, 2014, https://time.com/3628047/navy-laser-weapon-test-persian-gulf.

39. Louis A. Del Monte, *Nanoweapons: A Growing Threat to Humanity* (Lincoln NE: Potomac Books, 2017), 11–12.

40. Del Monte, *Nanoweapons*, xii.

41. Seyed Hashemipour, Ahmad Salmanogli, and N. Mohammadian, "The Effect of Atmosphere Disturbances on Laser Beam Propagation," *Key Engineering Materials* 500 (August 2012): 3–8, https://www.researchgate.net/publication/257932854_The_Effect_of_Atmosphere_Disturbances_on_Laser_BeamPropagation.

42. David Stoudt, "Myth or Reality: High-Energy Lasers Are Fair-Weather Weapons," Booz Allen Hamilton, 2017, https://www.boozallen.com/content/dam/boozallen_site/dig/pdf/publication/are-high-energy-lasers-fair-weather-weapons.pdf.

43. Stoudt, "Myth or Reality."

44. Patrick M. Shanahan, "2019 Missile Defense Review" (Washington DC: Office of the Secretary of Defense, January 2019), https://media.defense.gov/2019/Jan/17/2002080666/-1/-1/1/2019-MISSILE-DEFENSE-REVIEW.PDF.

45. Office of Corporate Communications, Naval Sea Systems Command, "Aegis Weapon System," U.S. Navy Fact File, https://www.navy.mil/navydata/fact_display.asp?cid=2100&tid=200&ct=2 (accessed July 7, 2019).

46. "Fact Sheet on U.S. Missile Defense Policy: A 'Phased, Adaptive Approach' for Missile Defense in Europe," Office of the Press Secretary, White House, September 17, 2009,

https://obamawhitehouse.archives.gov/the-press-office/fact-sheet-us-missile-defense
-policy-a-phased-adaptive-approach-missile-defense-eur.

47. "Next Generation Aegis Ballistic Missile Defense System Successfully Engages Medium Range Ballistic Missile Target," Lockheed Martin, February 6, 2017, https:// news.lockheedmartin.com/2017-02-06-Next-Generation-Aegis-Ballistic-Missile-Defense -System-Successfully-Engages-Medium-Range-Ballistic-Missile-Target.

48. Shanahan, "2019 Missile Defense Review."

49. "Lockheed Martin Receives $150 Million Contract to Deliver Integrated High Energy Laser Weapon Systems to U.S. Navy," Lockheed Martin news release, March 2, 2018, https:// news.lockheedmartin.com/2018-03-01-Lockheed-Martin-Receives-150-Million-Contract-to -Deliver-Integrated-High-Energy-Laser-Weapon-Systems-to-U-S-Navy. For the reference to the USS *Preble*, see David B. Larter, "When It Comes to Missile-Killing Lasers, the U.S. Navy Is Ready to Burn Its Ships," *Defense News*, May 22, 2019, https://www.defensenews.com/naval /2019/05/23/when-it-comes-to-missile-killing-lasers-the-us-navy-is-ready-to-burn-its-ships.

50. Sydney J. Freedberg Jr., "New Army Laser Could Kill Cruise Missiles," *Breaking Defense*, August 5, 2019, https://breakingdefense.com/2019/08/newest-army-laser-could -kill-cruise-missiles.

51. Freedberg, "New Army Laser."

52. Karen Yourish, "How Russia Hacked the Democrats in 2016," *New York Times*, July 13, 2018, https://www.nytimes.com/interactive/2018/07/13/us/politics/how-russia-hacked-the -2016-presidential-election.html; and Nicholas Fandos and Julian E. Barnes, "Republican-Led Review Backs Intelligence Findings on Russian Interference," *New York Times*, April 22, 2020, https://www.nytimes.com/2020/04/21/us/politics/russian-interference-senate -intelligence-report.html.

53. Andrew Tate, "Chinese Navy Trials Laser Weapon," Association of Old Crows, April 10, 2019, https://www.crows.org/news/446397/Chinese-navy-trials-laser-weapon.htm.

54. "China Tests Laser Weapon Similar to U.S. Navy Prototype," *Maritime Executive*, April 10, 2019, https://www.maritime-executive.com/article/china-tests-laser-weapon -similar-to-u-s-navy-prototype.

55. Bill Gertz, "DIA: China to Deploy ASAT Laser by 2020," *Washington Free Beacon*, February 15, 2019, https://freebeacon.com/national-security/dia-china-to-deploy-asat -laser-by-2020.

56. "China Tests Laser Weapon."

57. Defense Intelligence Agency, *China Military Power: Modernizing a Force to Fight and Win* (Washington DC: Defense Intelligence Agency, January 3, 2019), https://www.dia.mil /Portals/27/Documents/News/Military%20Power%20Publications/China_Military _Power_FINAL_5MB_20190103.pdf.

58. Edward Wong, "U.S. Versus China: A New Era of Great Power Competition, but Without Boundaries," *New York Times*, June 26, 2019, https://www.nytimes.com/2019/06 /26/world/asia/united-states-china-conflict.html.

59. "The People's Republic of China: U.S.-China Trade Facts," Office of the United States Trade Representative, https://ustr.gov/countries-regions/china-mongolia-taiwan /peoples-republic-china (accessed July 8, 2019).

60. Hannah Beech, "China's Sea Control Is a Done Deal, 'Short of War with the U.S.,'" *New York Times*, September 20, 2018, https://www.nytimes.com/2018/09/20/world/asia/south-china-sea-navy.html.

61. Frank Von Hippel and Thomas B. Cochran, "The Myth of the Soviet 'Killer' Laser," *New York Times*, August 19, 1989, https://www.nytimes.com/1989/08/19/opinion/the-myth-of-the-soviet-killer-laser.html.

62. Von Hippel and Cochran, "Soviet 'Killer' Laser."

63. William J. Broad, "The Secrets of Soviet Star Wars," *New York Times*, June 28, 1987, https://www.nytimes.com/1987/06/28/magazine/the-secrets-of-soviet-star-wars.html.

64. Von Hippel and Cochran, "Soviet 'Killer' Laser."

65. Broad, "Secrets of Soviet Star Wars."

66. Broad, "Secrets of Soviet Star Wars." Capitalization per original.

67. Missile Defense Project, "Russia Tests New Anti-Ballistic Missile Interceptor," *Missile Threat*, February 13, 2018, last modified June 15, 2018, https://missilethreat.csis.org/russia-tests-new-anti-ballistic-missile-interceptor/.

68. Tom O'Connor, "Russia's Military Has Laser Weapons That Can Take Out Enemies in Less Than a Second," *Newsweek*, March 12, 2018, https://www.newsweek.com/russia-military-laser-weapons-take-out-enemies-less-second-841091.

69. O'Connor, "Russia's Military."

70. "WATCH: Peresvet Combat Laser System Enter Test Duty in Russia (VIDEO)," RT, December 5, 2018, https://www.rt.com/news/445620-combat-laser-peresvet-deployment/.

71. "'Laser Weapons' to Define Russia's Military Potential in 21st Century—Putin," RT, May 17, 2019, https://www.rt.com/russia/459586-putin-laser-weapons-remarks.

5. Microwave Weapons

1. J. Mark Elwood, "Microwaves in the Cold War: The Moscow Embassy Study and Its Interpretation: Review of a Retrospective Cohort Study," *Environmental Health*, November 14, 2012, https://ehjournal.biomedcentral.com/articles/10.1186/1476-069X-11-85.

2. Sharon Weinberger, "The Secret History of Diplomats and Invisible Weapons," *Foreign Policy*, August 25, 2017, https://foreignpolicy.com/2017/08/25/the-secret-history-of-diplomats-and-invisible-weapons-russia-cuba.

3. Weinberger, "Secret History."

4. Weinberger, "Secret History."

5. P. Deshmukh et al., "Effect of Low-Level Microwave Radiation Exposure on Cognitive Function and Oxidative Stress in Rats," *Indian Journal of Biochemistry & Biophysics* 50, no. 2 (April 2013): 114–19, https://www.ncbi.nlm.nih.gov/pubmed/23720885.

6. Frances Robles and Kirk Semple, "'Health Attacks' on U.S. Diplomats in Cuba Baffle Both Countries," *New York Times*, August 11, 2017, https://www.nytimes.com/2017/08/11/world/americas/cuba-united-states-embassy-diplomats-illness.html.

7. Robles and Semple, "'Health Attacks.'"

8. Robles and Semple, "'Health Attacks.'"

9. Steven Lee Myers and Jane Perlez, "U.S. Diplomats Evacuated in China as Medical Mystery Grows," *New York Times*, June 6, 2018, https://www.nytimes.com/2018/06/06/world/asia/china-guangzhou-consulate-sonic-attack.html.

10. William J. Broad, "Microwave Weapons Are Prime Suspect in Ills of U.S. Embassy Workers," *New York Times*, September 1, 2018, https://www.nytimes.com/2018/09/01 /science/sonic-attack-cuba-microwave.html.

11. Broad, "Microwave Weapons."

12. Josh Lederman et al., "U.S. Officials Suspect Russia in Mystery 'Attacks' on Diplomats in Cuba, China," NBC News, September 11, 2018, https://www.nbcnews.com/news /latin-america/u-s-officials-suspect-russia-mystery-attacks-diplomats-cuba-china-n908141.

13. "Active Denial System FAQs," U.S. Department of Defense, Non-Lethal Weapons Program, https://jnlwp.defense.gov/About/Frequently-Asked-Questions/Active-Denial -System-FAQs (accessed July 20, 2019).

14. Mathieu Rabechault, "U.S. Military Unveils Non-Lethal Heat Ray Weapon," Phys .org, March 11, 2012, https://phys.org/news/2012-03-military-unveils-non-lethal-ray -weapon.html.

15. Rabechault, "U.S. Military Unveils."

16. "Award Details: Remote Personnel Incapacitation System," SBIR/STTR (Small Business Innovation Research/Small Business Technology Transfer) Search Database, 2004, https://www.sbir.gov/sbirsearch/detail/349489.

17. David Hambling, "Microwave Ray Gun Controls Crowds with Noise," *New Scientist*, July 3, 2008, https://www.newscientist.com/article/dn14250-microwave-ray-gun -controls-crowds-with-noise.

18. Hambling, "Microwave Ray Gun."

19. "Bioeffects of Selected Nonlethal Weapons," U.S. Army Intelligence and Security Command, Fort Meade MD, February 1998, http://thermoguy.com/wp-content/uploads /Bioeffects-of-Selected-Non-Lethal-Weapons.pdf; and David Hambling, "Secret Directed-Energy Tech Protecting the President? (Updated)," *Wired*, November 14, 2008, https:// www.wired.com/2008/11/presidents-secr.

20. James P. O'Loughlin and Diana L. Loree, "Method and Device for Implementing the Radio Frequency Hearing Effect," United States Patent 6,470,214 B1, October 22, 2002, https://patentimages.storage.googleapis.com/00/24/7c/4cf02f4210343e/US6470214.pdf.

21. Jeffry Pilcher, "Say It Again: Messages Are More Effective When Repeated," *Financial Brand*, https://thefinancialbrand.com/42323/advertising-marketing-messages-effective -frequency (accessed July 21, 2019).

22. Sarah O'Brien, "Consumers Cough Up $5,400 a Year on Impulse Purchases," CNBC, February 23, 2018, https://www.cnbc.com/2018/02/23/consumers-cough-up-5400-a -year-on-impulse-purchases.html.

23. Ronald Kessler, "U.S. Air Force Has Deployed 20 Missiles That Could Zap the Military Electronics of North Korea or Iran with Super Powerful Microwaves, Rendering Their Military Capabilities Virtually Useless with NO COLLATERAL DAMAGE," *Daily Mail*, May 16, 2019, https://www.dailymail.co.uk/news/article-7037549/Air-Force-deployed -20-missiles-fry-military-electronics-North-Korea-Iran.html.

24. Kessler, "U.S. Air Force."

25. "Boeing Non-kinetic Missile Records 1st Operational Test Flight," Boeing news release, Hill Air Force Base UT, October 22, 2012, https://boeing.mediaroom.com/2012 -10-22-Boeing-Non-kinetic-Missile-Records-1st-Operational-Test-Flight.

26. "Air Force Confirms Electromagnetic Pulse Weapon," CNN, May 25, 2015, https://www.cnn.com/videos/us/2015/05/25/orig-boeing-electromagnetic-pulse-weapon.cnn and https://www.youtube.com/watch?v=qo2mNFF6uGw.

27. Richard Pérez-Peña, "Iran Says It Seized Foreign Tanker, Escalating Regional Tensions," *New York Times*, July 18, 2019, https://www.nytimes.com/2019/07/18/world/middleeast/iran-oil-tanker.html; and David D. Kirkpatrick and Megan Specia, "Iran's Seizure of British Vessel Further Roils Gulf Region," *New York Times*, July 19, 2019, https://www.nytimes.com/2019/07/19/world/middleeast/iran-british-tanker-drone.html.

28. David Axe, "Commentary: Here's How the U.S. Navy Will Defeat Iran's Speedboats," Reuters, August 30, 2016, https://www.reuters.com/article/us-navy-iran-commentary/commentary-heres-how-the-u-s-navy-will-defeat-irans-speedboats-idUSKCN1151SB.

29. Nicholas D. Kristof, "How We Won the War," *New York Times*, September 6, 2002, https://www.nytimes.com/2002/09/06/opinion/how-we-won-the-war.html.

30. Steven Stashwick, "U.S. Navy Eyes More Combat Upgrades for Littoral Combat Ships," *Diplomat*, May 23, 2019, https://thediplomat.com/2019/05/us-navy-eyes-more-combat-upgrades-for-littoral-combat-ships.

31. Harold C. Hutchison, "U.S. Navy Ships Will Get Powerful Lasers to Zap Incoming Missiles," *Business Insider*, May 2, 2018, https://www.businessinsider.com/us-navy-ships-powerful-lasers-zap-incoming-missiles-2018-5.

32. "High-Power Microwaves and Lasers Defeat Multiple Drones during U.S. Army Exercise," Raytheon, March 12, 2018, http://raytheon.mediaroom.com/2018-03-20-High-power-microwaves-and-lasers-defeat-multiple-drones-during-US-Army-exercise.

33. "High-Power Microwaves."

34. Megan Eckstein, "Littoral Combat Ship Will Field Laser Weapon as Part of Lockheed Martin, Navy Test," *USNI News*, January 13, 2020, https://news.usni.org/2020/01/13/littoral-combat-ship-will-field-laser-weapon-as-part-of-lockheed-martin-navy-test.

35. Michael Schwirtz, "Russia and Cuba Sign Strategic Partnership," *New York Times*, January 30, 2009, https://www.nytimes.com/2009/01/31/world/europe/31russia.html.

36. Andrew Tieu, "New Microwave Gun Disables Missiles and Airplanes from 10 Kilometers Away," *Futurism*, June 16, 2015, https://futurism.com/new-microwave-gun-disables-missiles-and-airplanes-from-10-kilometers-away.

37. John Keller, "New Russian Directed-Energy Weapon Could Complicate U.S. Military Strategic Planning," *Military & Aerospace Electronics*, July 7, 2015, https://www.militaryaerospace.com/rf-analog/article/16714244/new-russian-directedenergy-weapon-could-complicate-us-military-strategic-planning.

38. Robert J. Capozzella, "High Power Microwaves on the Future Battlefield: Implications for U.S. Defense" (thesis, U.S. Air War College, February 17, 2010), http://www.ausairpower.net/PDF-A/Capozzella_awc_10.pdf.

39. Jeffrey Lin and P. W. Singer, "China's New Microwave Weapon Can Disable Missiles and Paralyze Tanks," *Popular Science*, January 26, 2017, https://www.popsci.com/china-microwave-weapon-electronic-warfare.

40. Richard D. Fisher Jr., "China's Progress with Directed Energy Weapons," Testimony before the U.S.-China Economic and Security Review Commission hearing "Chi-

na's Advanced Weapons," Washington DC, February 23, 2017, https://www.uscc.gov/sites/default/files/Fisher_Combined.pdf.

41. Fisher, "China's Progress."

42. Deng Xiaoci, "China Develops Weapon System Based on Microwave Radar Tech," *Global Times*, February 21, 2019, http://www.globaltimes.cn/content/1139703.shtml.

43. Clay Wilson, "High Altitude Electromagnetic Pulse (HEMP) and High Power Microwave (HPM) Devices: Threat Assessments" (Washington DC: Congressional Research Service, March 26, 2008), https://www.wired.com/images_blogs/dangerroom/files/Ebomb.pdf.

6. EMP Weapons

1. Nathan Yau, "A Day in the Life of Americans," FlowingData, https://flowingdata.com/2015/12/15/a-day-in-the-life-of-americans (accessed July 30, 2019).

2. Ariel Cohen, "Trump Moves to Protect America from Electromagnetic Pulse Attack," *Forbes*, April 5, 2019, https://www.forbes.com/sites/arielcohen/2019/04/05/whitehouse-prepares-to-face-emp-threat/#48677608e7e2.

3. "Surge and Lightning Damage to Electronics," StrikeCheck, https://strikecheck.com/2017/07/24/surge-lightning-damage-electronics (accessed July 30, 2019).

4. Kendra Pierre-Louis, "Heat Waves in the Age of Climate Change: Longer, More Frequent and More Dangerous," *New York Times*, July 18, 2019, https://www.nytimes.com/2019/07/18/climate/heatwave-climate-change.html.

5. "Are Power Surges Damaging Your Electronics?," State Farm, https://www.statefarm.com/simple-insights/residence/are-power-surges-damaging-your-electronics (accessed July 30, 2019).

6. "What Are Surges," NEMA Surge Protection Institute, https://www.nemasurge.org/history (accessed July 30, 2019).

7. "Michael Faraday (1791–1867)," BBC, https://www.bbc.co.uk/history/historic_figures/faraday_michael.shtml (accessed July 30, 2019).

8. Pierre-Louis, "Heat Waves."

9. "What Causes Power Surges and How to Protect Your Electronics," Prairie Electric, April 9, 2018, https://www.prairielectric.com/blog/what-causes-power-surges.

10. Robert Wilson, "Summary of Nuclear Physics Measurements," in *Trinity*, ed. K. T. Bainbridge (Los Alamos: Los Alamos Scientific Laboratory, May 1976), 53, https://fas.org/sgp/othergov/doe/lanl/docs1/00317133.pdf.

11. Wilson, "Summary."

12. *Operation HARDTACK Basic Effects Structures and Materiel* (Hollywood: Lookout Mountain Laboratory USAF, 1958), film, https://archive.org/details/OperationHARDTACK_BasicEffectsStructuresandMateriel1958.

13. Defense Atomic Support Agency, "Operation Hardtack Preliminary Report: Technical Summary of Military Effects, Programs 1–9," Report ADA369152 (Sandia Base, Albuquerque: Defense Atomic Support Agency, September 23, 1959), https://apps.dtic.mil/dtic/tr/fulltext/u2/a369152.pdf.

14. "9 July 1962, 'Starfish Prime,' Outer Space," Comprehensive Nuclear-Test-Ban Treaty Organization, https://www.ctbto.org/specials/testing-times/9-july-1962starfish-prime-outer-space (accessed August 2, 2019).

15. "9 July 1962."

16. David S. F. Portree, "Starfish and Apollo (1962)," *Wired*, March 21, 2012, https://www.wired.com/2012/03/starfishandapollo-1962.

17. Phil Plait, "The 50th Anniversary of Starfish Prime: The Nuke that Shook the World," *Discover*, July 9, 2012, https://www.discovermagazine.com/the-sciences/the-50th-anniversary-of-starfish-prime-the-nuke-that-shook-the-world.

18. Anne Marie Helmenstine, "Starfish Prime: The Largest Nuclear Test in Space," ThoughtCo., June 2, 2019, https://www.thoughtco.com/starfish-prime-nuclear-test-4151202.

19. Howard Seguine, "U.S.-Russian Meeting—HEMP Effects on National Power Grid & Telecommunications," Nuclear Weapon Archive, February 17, 1995, http://nuclearweaponarchive.org/News/Loborev.txt.

20. "Cuban Missile Crisis," History, January 4, 2010, https://www.history.com/topics/cold-war/cuban-missile-crisis.

21. "Cuban Missile Crisis: Tuesday, October 25, 1962," U.S. Department of State, May 27, 2010, https://web.archive.org/web/20100527171609/http://future.state.gov/educators/slideshow/cuba/cuba2.html.

22. "President John F. Kennedy on the Cuban Missile Crisis," The History Place: Great Speeches Collection, October 22, 1962, http://www.historyplace.com/speeches/jfk-cuban.htm.

23. Gilbert King, "Going Nuclear over the Pacific," *Smithsonian*, August 15, 2012, https://www.smithsonianmag.com/history/going-nuclear-over-the-pacific-24428997.

24. Stephen Younger et al., "Scientific Collaborations between Los Alamos and Arzamas-16 Using Explosive-Driven Flux Compression Generators," *Los Alamos Science* 24 (1996): 48–71, https://fas.org/sgp/othergov/doe/lanl/pubs/00326620.pdf.

25. *Encyclopaedia Britannica*, s.v. "pinch effect," https://www.britannica.com/science/pinch-effect (accessed August 3, 2019).

26. Erwin Marx, "Versuche über die Prüfung von Isolatoren mit Spannungsstößen" [Experiments on the testing of insulators using high-voltage pulses], *Elektrotechnische Zeitschrift* 25 (1924): 652–54.

27. "How Does the Z Machine Work?," Sandia National Laboratories, https://www.sandia.gov/z-machine/about_z/how-z-works.html (accessed August 7, 2019).

28. "Sample Records for Solid-State Marx Generator," Report 15, Science.gov, https://www.science.gov/topicpages/s/solid-state+marx+generator.html (accessed August 3, 2019).

29. Tom Harris, "How E-Bombs Work," How Stuff Works, March 23, 2003, https://science.howstuffworks.com/e-bomb3.htm.

30. Google search using the phrase "how dangerous are EMP attacks," www.google.com, August 4, 2019.

31. Paul Bedard, "Military Warns EMP Attack Could Wipe Out America, 'Democracy, World Order,'" *Washington Examiner*, November 30, 2018, https://www.washingtonexaminer.com/washington-secrets/military-warns-emp-attack-could-wipe-out-america-democracy-world-order; and David Stuckenberg, James Woolsey, and Douglas DeMaio, "Electromagnetic Defense Task Force (EDTF) 2018 Report," LeMay Paper No. 2 (Maxwell Air Force Base AL: Air University Press, November 2018), https://media.defense.gov/2018/Nov/28/2002067172/-1/-1/0/LP_0002_DEMAIO_ELECTROMAGNETIC_DEFENSE_TASK_FORCE.PDF.

32. U.S. House Committee on Armed Services, "Threat Posed by Electromagnetic Pulse (EMP) Attack," H.A.S.C. No. 110-156 (Washington DC: U.S. Government Printing Office, 2009), 8, https://www.govinfo.gov/content/pkg/CHRG-110hhrg45133/pdf/CHRG-110hhrg45133.pdf.

33. Anne Marie Helmenstine, "Gamma Radiation Definition," ThoughtCo., March 4, 2019, https://www.thoughtco.com/definition-of-gamma-radiation-604476.

34. Conrad L. Longmire, "Justification and Verification of High-Altitude EMP Theory, Part 1" (Livermore CA: Lawrence Livermore National Laboratory, June 1986), http://ece-research.unm.edu/summa/notes/TheoreticalPDFs/TN368.pdf.

35. "Terminal High Altitude Area Defense (THAAD)," U.S. Department of Defense, Missile Defense Agency, https://www.mda.mil/system/thaad.html (accessed August 5, 2019).

36. William R. Graham and Peter Vincent Pry, "North Korea Nuclear EMP Attack: An Existential Threat," Statement for the Record to the U.S. House of Representatives Committee on Homeland Security hearing "Empty Threat or Serious Danger: Assessing North Korea's Risk to the Homeland," October 12, 2017, https://docs.house.gov/meetings/HM/HM09/20171012/106467/HHRG-115-HM09-Wstate-PryP-20171012.pdf.

37. Tom O'Connor, "North Korea Threatens Nuclear War with U.S., but 'Loves Peace More Than Anyone Else,'" Newsweek, November 13, 2017, https://www.newsweek.com/north-korea-threatens-nuclear-war-us-loves-peace-more-anyone-else-709740.

38. James Griffiths, Joshua Berlinger, and Sheena McKenzie, "Iranian Leader Announces Partial Withdrawal from Nuclear Deal," CNN, May 8, 2019, https://www.cnn.com/2019/05/08/middleeast/iran-nuclear-deal-intl/index.html.

39. Geoff Brumfiel, "Iran's Uranium Enrichment Breaks Nuclear Deal Limit: Here's What That Means," All Things Considered, NPR, July 7, 2019, https://www.npr.org/2019/07/07/738902822/irans-uranium-enrichment-breaks-nuclear-deal-limit-here-s-what-that-means.

40. Dennis Overbye, "Living with a Star," New York Times, February 4, 2015, https://www.nytimes.com/2015/02/05/science/living-with-a-star.html.

41. Donald J. Trump, "Executive Order on Coordinating National Resilience to Electromagnetic Pulses," White House, March 26, 2019, https://www.whitehouse.gov/presidential-actions/executive-order-coordinating-national-resilience-electromagnetic-pulses.

42. Trump, "Executive Order."

7. Cyberspace Weapons

1. Leinz Vales, "The Moment CNN Projected Donald Trump Is President-Elect," CNN, November 9, 2016, https://www.cnn.com/2016/11/09/politics/moment-cnn-projects-trump-u-s-president-cnntv/index.html.

2. Nate Silver, "The Comey Letter Probably Cost Clinton the Election," FiveThirtyEight, May 3, 2017, https://fivethirtyeight.com/features/the-comey-letter-probably-cost-clinton-the-election.

3. Silver, "Comey Letter."

4. Office of the Director of National Intelligence, "Background to 'Assessing Russian Activities and Intentions in Recent US Elections': The Analytic Process and Cyber Incident Attribution" (Washington DC: Office of the Director of National Intelligence, January 6, 2017), 1, https://www.dni.gov/files/documents/ICA_2017_01.pdf.

5. Jonathan Masters, "Russia, Trump, and the 2016 U.S. Election," Council on Foreign Relations, February 26, 2018, https://www.cfr.org/backgrounder/russia-trump-and-2016-us-election.

6. United States of America v. Internet Research Agency LLC et al., Criminal No. (18 U.S.C. §§ 2, 371, 1349, 1028A), in the United States District Court for the District of Columbia, February 16, 2018, https://www.justice.gov/file/1035477/download.

7. Chris Megerian, "Bitcoin, Malware and 'Spearphishing' Helped Russian Agents Hack Democratic Party Computers in 2016 Election," *Los Angeles Times*, July 15, 2018, https://www.latimes.com/politics/la-na-pol-russian-hacking-ticktock-20180715-story.html.

8. Megerian, "Bitcoin, Malware."

9. *Merriam-Webster*, s.v. "cryptocurrency," https://www.merriam-webster.com/dictionary/cryptocurrency (accessed September 6, 2019).

10. Maria Tsvetkova and Roman Anin, "'We Are Hardly Surviving': As Oil and the Ruble Drop, Ordinary Russians Face Growing List of Problems," *Financial Post*, January 2, 2015, https://business.financialpost.com/news/economy/we-are-hardly-surviving-as-oil-and-the-ruble-drop-ordinary-russians-face-growing-list-of-problems.

11. Nathaniel Popper and Matthew Rosenberg, "How Russian Spies Hid behind Bitcoin in Hacking Campaign," *New York Times*, July 13, 2018, https://www.nytimes.com/2018/07/13/technology/bitcoin-russian-hacking.html.

12. Popper and Rosenberg, "How Russian Spies."

13. United States of America v. Internet Research Agency LLC et al.

14. Jonathan Masters, "Russia, Trump, and the 2016 U.S. Election," Council on Foreign Relations, February 26, 2018, https://www.cfr.org/backgrounder/russia-trump-and-2016-us-election.

15. Office of the Director of National Intelligence, "Background to 'Assessing,'" ii.

16. Eric Lipton, David E. Sanger, and Scott Shane, "The Perfect Weapon: How Russian Cyberpower Invaded the U.S.," *New York Times*, December 13, 2016, https://www.nytimes.com/2016/12/13/us/politics/russia-hack-election-dnc.html.

17. Lipton, Sanger, and Shane, "Perfect Weapon."

18. Cory Bennett and Martin Matishak, "Democrats: Did Americans Help Russia Hack the Election?," *Politico*, July 5, 2017, https://www.politico.com/story/2017/07/05/who-helped-russia-hack-us-election-240210.

19. Lipton, Sanger, and Shane, "Perfect Weapon."

20. Kim Zetter, "An Unprecedented Look at Stuxnet, the World's First Digital Weapon," *Wired*, November 3, 2014, https://www.wired.com/2014/11/countdown-to-zero-day-stuxnet.

21. Zetter, "Unprecedented Look."

22. Eugene Kaspersky, "The Man Who Found Stuxnet—Sergey Ulasen in the Spotlight," November 2, 2011, https://eugene.kaspersky.com/2011/11/02/the-man-who-found-stuxnet-sergey-ulasen-in-the-spotlight.

23. Kaspersky, "Man Who Found."

24. Robert Lipovsky, "Seven Years after Stuxnet: Industrial Systems Security Once Again in the Spotlight," WeLiveSecurity, June 16, 2017, https://www.welivesecurity.com/2017/06/16/seven-years-stuxnet-industrial-systems-security-spotlight.

25. "What Is Stuxnet?," McAfee, https://www.mcafee.com/enterprise/en-us/security-awareness/ransomware/what-is-stuxnet.html (accessed August 16, 2019).

26. Michael Holloway, "Stuxnet Worm Attack on Iranian Nuclear Facilities," submitted as coursework for PH241 "Introduction to Nuclear Energy," Stanford University, July 16, 2015, http://large.stanford.edu/courses/2015/ph241/holloway1.

27. William J. Broad, John Markoff, and David E. Sanger, "Israeli Test on Worm Called Crucial in Iran Nuclear Delay," *New York Times*, January 15, 2011, https://www.nytimes.com/2011/01/16/world/middleeast/16stuxnet.html.

28. Broad, Markoff, and Sanger, "Israeli Test."

29. Lisa Ferdinando, "Cybercom to Elevate to Combatant Command," DOD *News*, May 3, 2018, https://www.defense.gov/Newsroom/News/Article/Article/1511959/cybercom-to-elevate-to-combatant-command.

30. "U.S. Cyber Command," https://www.cybercom.mil/About/Mission-and-Vision/ (accessed August 24, 2019).

31. "Summary: Department of Defense Cyber Strategy, 2018" (Washington DC: Department of Defense, 2018), https://media.defense.gov/2018/Sep/18/2002041658/-1/-1/1/CYBER_STRATEGY_SUMMARY_FINAL.PDF. Emphasis in original.

32. Doug G. Ware, "NATO Officially Recognizes Cyberspace as Domain for War," UPI, June 14, 2016, https://www.upi.com/Top_News/World-News/2016/06/14/NATO-officially-recognizes-cyberspace-as-domain-for-war/2271465941545.

33. Ware, "NATO Officially Recognizes."

34. Google search using the phrase "electronic warfare," www.google.com, August 20, 2019.

35. "The Evolution of Electronic Warfare: A Timeline," Army Technology, June 7, 2018, https://www.army-technology.com/features/evolution-electronic-warfare-timeline.

36. "World War II: The Battle of Britain," Ducksters Education Site, https://www.ducksters.com/history/world_war_ii/battle_of_britain.php (accessed August 19, 2019).

37. "World War II."

38. C. N. Trueman, "The Radar and the Battle of Britain," History Learning Site, April 21, 2015, https://www.historylearningsite.co.uk/world-war-two/world-war-two-in-western-europe/battle-of-britain/the-radar-and-the-battle-of-britain/.

39. "World War II."

40. Trueman, "Radar."

41. "World War II."

42. Gen. Peter Pace, "Joint Publication 3-13.1: Electronic Warfare" (Washington DC: Joint Chiefs of Staff, Department of Defense, February 8, 2012), https://fas.org/irp/doddir/dod/jp3-13-1.pdf.

43. U.S. Marine Corps, "Section VII. Electronic Warfare," in MCWP 3-15.5: MAGTF *Antiarmor Operations* (Washington DC: Department of the Navy, February 2000), https://www.globalsecurity.org/military/library/policy/usmc/mcwp/3-15-5/cdraft_ch4-7.pdf.

44. Madison Creery, "The Russian Edge in Electronic Warfare," *Georgetown Security Studies Review*, June 26, 2019, https://georgetownsecuritystudiesreview.org/2019/06/26/the-russian-edge-in-electronic-warfare/#_edn2.

45. Jared Keller, "After Experiencing Russian Jamming up Close in Syria, the Pentagon Is Scrambling to Catch Up," *Business Insider*, June 3, 2019, https://www.businessinsider.com/pentagon-focus-on-electronic-warfare-after-russian-jamming-in-syria-2019-6.

46. Creery, "Russian Edge."

47. Creery, "Russian Edge."

48. Keller, "After Experiencing."

49. Creery, "Russian Edge."

50. David Deming, "Do Extraordinary Claims Require Extraordinary Evidence?," *Philosophia* 44, no. 4 (2016): 1319–31, https://www.ncbi.nlm.nih.gov/pmc/articles/PMC6099700.

51. Hans Christian Ørsted, *Selected Scientific Works of Hans Christian Ørsted*, trans. and ed. Karen Jelved, Andrew D. Jackson, and Ole Knudsen (Princeton NJ: Princeton University Press, 1998), 421–45.

52. Sam LaGrone, "U.S. Official: Russia Installed System in Crimea to Snoop on U.S. Destroyers, Jam Communications," *USNI News*, May 1, 2017, https://news.usni.org/2017/05/01/official-russia-installs-system-crimea-snoop-u-s-destroyers-jam-communications.

53. Keller, "After Experiencing."

54. John R. Hoehn, "U.S. Military Electronic Warfare Investment Funding: Background and Issues for Congress" (Washington DC: Congressional Research Service, June 6, 2019), https://fas.org/sgp/crs/natsec/R45756.pdf.

55. Garrett M. Graff, "China's 5 Steps for Recruiting Spies," *Wired*, October 31, 2018, https://www.wired.com/story/china-spy-recruitment-us.

56. Graff, "China's 5 Steps."

57. John C. Demers, Statement before U.S. Senate Committee on the Judiciary hearing "China's Non-Traditional Espionage against the United States: The Threat and Potential Policy Responses," December 12, 2018, 4, https://www.justice.gov/sites/default/files/testimonies/witnesses/attachments/2018/12/18/12-05-2018_john_c._demers_testimony_re_china_non-traditional_espionage_against_the_united_states_the_threat_and_potential_policy_responses.pdf.

58. Demers, Statement.

59. Demers, Statement, 5.

60. Demers, Statement.

61. Lyu Jinghua, "What Are China's Cyber Capabilities and Intentions?," Carnegie Endowment for International Peace, April 1, 2019, https://carnegieendowment.org/2019/04/01/what-are-china-s-cyber-capabilities-and-intentions-pub-78734.

62. Bruce Sussman, "Cyber Attack Motivations: Russia vs. China," Secure World, June 3, 2019, https://www.secureworldexpo.com/industry-news/why-russia-hacks-why-china-hacks.

63. Office of the Secretary of Defense, "Annual Report to Congress: Military and Security Developments Involving the People's Republic of China 2019" (Washington DC: Department of Defense, May 2, 2019), https://media.defense.gov/2019/May/02/2002127082/-1/-1/1/2019%20CHINA%20MILITARY%20POWER%20REPORT%20(1).PDF.

64. Zak Doffman, "'National Security Threat' as Chinese Hackers Are 'Allowed' to Target U.S. Businesses," *Forbes*, April 13, 2019, https://www.forbes.com/sites/zakdoffman

/2019/04/13/u-s-businesses-allowing-chinese-government-hackers-to-steal-american
-secrets/#394ea3d8a4d4.

65. David E. Sanger, David Barboza, and Nicole Perlroth, "Chinese Army Unit Is
Seen as Tied to Hacking against U.S.," *New York Times*, February 18, 2013, https://
www.nytimes.com/2013/02/19/technology/chinas-army-is-seen-as-tied-to-hacking
-against-us.html.

66. David Sanger, Nicole Perloth, and Scott Shane, "How Chinese Spies Got the N.S.A.'s
Hacking Tools, and Used Them for Attacks," *New York Times*, May 7, 2019, https://www
.nytimes.com/2019/05/06/us/politics/china-hacking-cyber.html.

67. Mark Pomerleau, "New Report Explains How China Thinks about Information
Warfare," *c4isrnet*, May 3, 2019, https://www.c4isrnet.com/c2-comms/2019/05/03
/new-report-explains-how-china-thinks-about-information-warfare.

68. Office of the Secretary of Defense, "Annual Report," 64.

69. Ashton B. Carter, "Autonomy in Weapon Systems," Department of Defense Directive No. 3000.09, November 21, 2012; incorporating change 1, May 8, 2017, https://fas.org
/irp/doddir/dod/d3000_09.pdf.

8. Directed-Energy Countermeasures

1. Eric Jackson, "Sun Tzu's 31 Best Pieces of Leadership Advice," *Forbes*, May 23, 2014,
https://www.forbes.com/sites/ericjackson/2014/05/23/sun-tzus-33-best-pieces-of
-leadership-advice/#64327d9f5e5e.

2. Office of Technology Assessment, "Countermeasures to Weapons," in *Strategic Defenses:
Two Reports by the Office of Technology Assessment* (Princeton NJ: Princeton University
Press, July 14, 2014), https://books.google.com/books?id=jsH_AwAAQBAJ&pg=PA172
#v=onepage&q&f=false.

3. Stephen Chen, "U.S. Lasers? PLA Preparing to Raise Its Deflector Shields," *South
China Morning Post*, March 10, 2014, https://www.scmp.com/news/china/article/1444732
/us-lasers-pla-preparing-raise-its-deflector-shields.

4. Chen, "U.S. Lasers?"

5. Megan Trimble, "U.S. Navy Awards $150M Contract to Develop Lasers," *U.S. News
& World Report*, March 2, 2018, https://www.usnews.com/news/national-news/articles
/2018-03-02/us-navy-awards-lockheed-martin-150m-contract-to-develop-laser-system.

6. Chen, "U.S. Lasers?"

7. Wilson, "High Altitude."

8. Wilson, "High Altitude."

9. Wilson, "High Altitude."

10. Wilson, "High Altitude."

11. "High-Power Microwave (HPM) / E-Bomb," Global Security, https://www
.globalsecurity.org/military/systems/munitions/hpm.htm (accessed August 30, 2019);
and Wilson, "High Altitude."

12. Cohen, "Trump Moves."

13. Kashmira Gander, "Software Giant Microsoft Has Revealed It Is Paying a Hacker over
$100,000 to Find Security Holes in Its Products," *Independent*, October 10, 2013, https://

www.independent.co.uk/life-style/gadgets-and-tech/news/microsoft-pays-out-100000
-to-hacker-who-exposed-windows-security-flaws-8871042.html.

14. Davey Winder, "Microsoft Confirms It Has Paid $4.4M to Hackers," *Forbes*, August
6, 2019, https://www.forbes.com/sites/daveywinder/2019/08/06/microsoft-confirms-it
-has-paid-44m-to-hackers/#f4dfeaf6c218.

15. Winder, "Microsoft Confirms."

16. Nicole Perlroth and Scott Shane, "In Baltimore and Beyond, a Stolen N.S.A. Tool
Wreaks Havoc," *New York Times*, May 25, 2019, https://www.nytimes.com/2019/05/25
/us/nsa-hacking-tool-baltimore.html.

17. Maggie Hassan, "Bipartisan Hassan-Portman Hack DHS Act Unanimously Passes
Senate," U.S. Senate, April 18, 2018, https://www.hassan.senate.gov/news/press-releases
/bipartisan-hassan-portman-hack-dhs-act-unanimously-passes-senate.

18. Lauren C. Williams, "Lessons Learned from DOD's Bug Bounty Programs," *Federal Computer Week*, October 5, 2018, https://fcw.com/articles/2018/10/05/dds-bug
-bounty-williams.aspx.

19. Williams, "Lessons Learned."

20. "Electronic Warfare," Lockheed Martin, https://www.lockheedmartin.com/en
-us/capabilities/electronic-warfare.html (accessed September 2, 2019); "Electronic Warfare," Northrop Grumman, https://www.northropgrumman.com/sea/electronic-warfare
(accessed May 20, 2020); and "Electronic Warfare," Raytheon, https://www.raytheon.com
/capabilities/ew (accessed September 2, 2019).

21. "Lockheed Martin Awarded $184 Million to Continue Providing the U.S. Navy
with Electronic Warfare Systems," Lockheed Martin, February 22, 2019, https://news
.lockheedmartin.com/2019-02-11-Lockheed-Martin-Awarded-184-Million-to-Continue
-Providing-the.

22. Justin Rohrlich and Dave Gershgorn, "The U.S. Navy Wants Electronic Weapons
That 'Go to 11,'" *Quartz*, November 29, 2018, https://qz.com/1478054/the-us-navy-wants
-electronic-warfare-equipment-that-goes-to-11.

23. Alan Boyle, "Boeing and the U.S. Navy Turn Growler Jets into Remote-Controlled
Drones as 'Force Multiplier,'" *GeekWire*, February 4, 2020, https://www.geekwire.com
/2020/boeing-u-s-navy-turn-growler-jets-remote-controlled-drones-force-multiplier/.

24. Sydney J. Freedberg Jr., "Digital Arsenal: Army Inches forward on Electronic Warfare," *Breaking Defense*, August 9, 2019, https://breakingdefense.com/2019/08/army-inches
-forward-on-electronic-warfare.

25. Freedberg, "Digital Arsenal."

26. Freedberg, "Digital Arsenal."

27. Freedberg, "Digital Arsenal."

28. Mark Pomerleau, "What DoD's Weapon Tester Said about Army Electronic Warfare," *C4ISRNET*, February 7, 2020, https://www.c4isrnet.com/electronic-warfare/2020
/02/07/what-dods-weapon-tester-said-about-army-electronic-warfare/.

29. Valerie Insinna, "Air Force to Wrap Up Electronic Warfare Study by January," *C4IS-
RNET*, November 14, 2018, https://www.c4isrnet.com/electronic-warfare/2018/11/14/air
-force-to-wrap-up-electronic-warfare-study-by-january.

30. John Keller, "Navy Surveys U.S. Prime Defense Contractors for EA-18G Jet Electronic Warfare (ew) Low-Band Jammer Pod," *Military & Aerospace Electronics,* May 16, 2019, https://www.militaryaerospace.com/sensors/article/14033399/navy-surveys-us-prime-defense-contractors-for-ea18g-jet-electronic-warfare-ew-lowband-jammer-pod.

31. "Air Force Announces Electronic Warfare, Electromagnetic Spectrum Superiority Enterprise Capability Collaboration Team Results," Secretary of the Air Force Public Affairs, April 16, 2019, https://www.af.mil/News/Article-Display/Article/1816024/air-force-announces-electronic-warfare-electromagnetic-spectrum-superiority-ent.

9. Force Fields

1. Brian J. Tillotson, "Method and System for Shockwave Attenuation via Electromagnetic Arc," United States Patent 8,981,261, March 17, 2015, http://patft.uspto.gov/netacgi/nph-Parser?Sect1=PTO1&Sect2=HITOFF&d=PALL&p=1&u=%2Fnetahtml%2FPTO%2Fsrchnum.htm&r=1&f=G&l=50&s1=8981261.PN.&OS=PN/8981261&RS=PN/8981261.

2. HS Vortex, "IED Awareness: The Blast Zone," *Medium,* November 30, 2015, https://medium.com/homeland-security/ied-awareness-the-blast-zone-be8efd4c837c.

3. Myles Gough, "Boeing Has Patented a Plasma 'Force Field' to Protect against Shock Waves," Science Alert, March 30, 2015, https://www.sciencealert.com/boeing-has-patented-a-plasma-force-field-to-protect-against-shock-waves.

4. Ehsan Vadiee, "Microwave Triggered Laser Ionization of Air" (thesis, University of New Mexico, September 3, 2013), https://pdfs.semanticscholar.org/dfcc/250c98d084c4ad9ae117f30d8e46e34e1724.pdf.

5. "Active Protection Systems," General Dynamic, https://www.gd-ots.com/protection-systems/active-protection-systems/ (accessed May 20, 2020).

6. "Hard-Kill Active Defense Solutions—Manufacturer: KBP," *Defense Update,* 2004, updated October 12, 2005, https://web.archive.org/web/20100414022528/http://defense-update.com/products/d/drozd-2.htm.

7. Tom J. Meyer, "Active Protective Systems: Impregnable Armor or Simply Enhanced Survivability?," *Armor,* May–June 1998, https://fas.org/man/dod-101/sys/land/docs/3aps98.pdf.

8. "Analysis Russian Afganit Active Protection System Is Able to Intercept Uranium Tank Ammunition TASS 11012163," Army Recognition, December 10, 2016, https://www.armyrecognition.com/weapons_defence_industry_military_technology_uk/analysis_russian_afganit_active_protection_system_is_able_to_intercept_uranium_tank_ammunition_tass_11012163.html.

9. "Iron Fist (Israel)," Missile Defense Advocacy Alliance, https://missiledefenseadvocacy.org/missile-defense-systems-2/allied-air-and-missile-defense-systems/allied-intercept-systems-coming-soon/iron-fist-israel (accessed September 12, 2019); and "Trophy Active Protection System," *Defense Update,* April 10, 2007, https://defense-update.com/20070410_trophy-2.html.

10. Tal Inbar, "Integrating 'Trophy' and 'Iron Fist,'" *Israel Defense,* September 14, 2019, https://www.israeldefense.co.il/en/content/integrating-%E2%80%9Ctrophy%E2%80%9D-and-%E2%80%9Ciron-fist%E2%80%9D.

11. Sydney J. Freedberg Jr., "300 Shots: Rafael Readies Trophy Lite for U.S. Stryker," *Breaking Defense*, September 25, 2018, https://breakingdefense.com/2018/09/300-shots-rafael-readies-trophy-lite-for-us-stryker.

12. Joseph Trevithick, "The Warzone: Army Hits Setbacks in Search for Active Protection System to Go on Its Strykers," *Drive*, June 10, 2019, https://www.thedrive.com/the-war-zone/28454/army-hits-setbacks-in-search-for-active-protection-system-to-go-on-its-strykers.

13. *Lexico*, s.v. "radiation," https://www.lexico.com/en/definition/radiation (accessed September 10, 2019).

14. Jason Perez, ed., "Analog Missions: Why Space Radiation Matters," NASA, https://www.nasa.gov/analogs/nsrl/why-space-radiation-matters (accessed September 10, 2019).

15. Perez, "Why Space Radiation Matters."

16. Antonella Del Rosso, "World-Record Current in the MgB_2 Superconductor," CERN *Bulletin*, April 14, 2014, https://cds.cern.ch/journal/CERNBulletin/2014/16/News%20Articles/1693853?ln=en.

17. Peter Dockrill, "Scientists Are Developing a Shield to Protect Astronauts from Cosmic Radiation," Science Alert, August 6, 2015, https://www.sciencealert.com/scientists-are-developing-a-magnetic-shield-to-protect-astronauts-from-cosmic-radiation.

18. Dockrill, "Scientists Are Developing."

19. *Encyclopedia Britannica*, s.v. "fundamental interaction," https://www.britannica.com/science/fundamental-interaction (accessed September 12, 2019).

10. Autonomous Directed-Energy Weapons

1. "War Quotes: Sun Tzu," Brainy Quote, https://www.brainyquote.com/quotes/sun_tzu_751632?src=t_war (accessed September 24, 2019).

2. Carter, "Autonomy in Weapon Systems."

3. Kelley M. Sayler, "Defense Primer: U.S. Policy on Lethal Autonomous Weapon Systems" (Washington DC: Congressional Research Service, March 27, 2019, updated December 19, 2019), https://fas.org/sgp/crs/natsec/IF11150.pdf.

4. Vincent C. Müller and Nick Bostrom, "Future Progress in Artificial Intelligence: A Survey of Expert Opinion," Oxford University, 2014, 1, https://nickbostrom.com/papers/survey.pdf.

5. Ariel Conn, "The Risks Posed by Lethal Autonomous Weapons," Future of Life Institute, September 4, 2018, https://futureoflife.org/2018/09/04/the-risks-posed-by-lethal-autonomous-weapons/?cn-reloaded=1.

6. International Committee of the Red Cross, "Ethics and Autonomous Weapon Systems: An Ethical Basis for Human Control?," *Arms Control Today*, July/August 2018, https://www.armscontrol.org/act/2018-07/features/document-ethics-autonomous-weapon-systems-ethical-basis-human-control.

7. Carter, "Autonomy in Weapon Systems."

8. "X-47B Unmanned Combat Air System (UCAS)," Naval Technology, https://www.naval-technology.com/projects/x-47b-unmanned-combat-air-system-carrier-ucas (accessed September 18, 2019).

9. Daniel Cooper, "The Navy's Unmanned Drone Project Gets Pushed Back a Year," *Engadget*, February 5, 2015, https://www.engadget.com/2015/02/05/drone-project-pushed -back-to-2016.

10. Carter, "Autonomy in Weapon Systems."

11. Amitai Etzioni and Oren Etzioni, "Pros and Cons of Autonomous Weapons Systems," *Military Review*, May–June 2017, https://www.armyupress.army.mil/Journals/Military -Review/English-Edition-Archives/May-June-2017/Pros-and-Cons-of-Autonomous -Weapons-Systems.

12. Geoff Brumfiel, "What We Know about the Attack on Saudi Oil Facilities," NPR, September 19, 2019, https://www.npr.org/2019/09/19/762065119/what-we-know-about -the-attack-on-saudi-oil-facilities.

13. Michael Safi and Julian Borger, "How Did Oil Attack Breach Saudi Defences and What Will Happen Next?," *Guardian*, September 18, 2019, https://www.theguardian.com /world/2019/sep/19/how-did-attack-breach-saudi-defences-and-what-will-happen-next.

14. Patrick Tucker, "Russia to the United Nations: Don't Try to Stop Us from Building Killer Robots," *Defense One*, November 21, 2017, https://www.defenseone.com/technology /2017/11/russia-united-nations-dont-try-stop-us-building-killer-robots/142734.

15. Siobhán O'Grady, "What Is the Russian s-400 Air Defense System, and Why Is the U.S. Upset Turkey Bought It?," *Washington Post*, July 17, 2019, https://www.washingtonpost .com/world/2019/07/12/what-is-russian-s-air-defense-system-why-is-us-upset-turkey -bought-it.

16. Dave Majumdar, "Can Russia's s-300 and s-400 Beat America's Stealth Fighters in Battle?," *National Interest*, August 2016, https://nationalinterest.org/blog/buzz/can-russias -s-300-and-s-400-beat-americas-stealth-fighters-battle-67707.

17. Brian Wang, "U.S. Defense Pushing for Megawatt Class Lasers," Next Big Future, April 25, 2018, https://www.nextbigfuture.com/2018/04/us-defense-pushing-for-megawatt -class-lasers.html.

18. Wang, "U.S. Defense."

19. Brian Wang, "U.S. Spending $300 Million and Rising to $1 Billion by 2020 to Begin Combat Laser Rollout," Next Big Future, February 15, 2018, https://www.nextbigfuture .com/2018/02/us-spending-300-million-and-rising-to-1-billion-by-2020-to-begin-combat -laser-rollout.html.

11. The Weaponization of Space

1. "Milstar Satellite Communications System," U.S. Air Force Space Command, Public Affairs Office, November 23, 2015, https://www.af.mil/About-Us/Fact-Sheets/Display /Article/104563/milstar-satellite-communications-system.

2. Israel Vargas, "Attacking Satellites Is Increasingly Attractive—and Dangerous," *Economist*, July 18, 2019, https://www.economist.com/briefing/2019/07/18/attacking-satellites -is-increasingly-attractive-and-dangerous.

3. Vargas, "Attacking Satellites."

4. "Defense Support Program Satellites," U.S. Air Force Space Command, Public Affairs Office, November 23, 2015, https://www.af.mil/About-Us/Fact-Sheets/Display/Article /104611/defense-support-program-satellites.

5. "Revelation 8:7," New King James Version, https://biblehub.com/revelation/8-7.htm (accessed September 28, 2019).

6. "Radiation Hardened SOI CMOS Technology," Honeywell Aerospace, https://aerospace.honeywell.com/content/dam/aero/en-us/documents/learn/products/sensors/brochures/RadiationHardenedSOICMOSTechnology-bro.pdf (accessed May 20, 2020).

7. "Milstar Satellite."

8. "The Global Positioning System: What Is GPS?," https://www.gps.gov/systems/gps (accessed September 30, 2019).

9. Benjamin Elisha Sawe, "How Many Types of Satellites Are There?," *World Atlas*, last updated April 25, 2017, https://www.worldatlas.com/articles/how-many-types-of-satellites-are-there.html.

10. "Treaty on Principles Governing the Activities of States in the Exploration and Use of Outer Space, Including the Moon and Other Celestial Bodies," United Nations Office for Disarmament Affairs, October 10, 1967, http://disarmament.un.org/treaties/t/outer_space.

11. "Treaty on Principles."

12. Johnny Wood, "The Countries with the Most Satellites in Space," World Economic Forum, May 4, 2019, https://www.weforum.org/agenda/2019/03/chart-of-the-day-the-countries-with-the-most-satellites-in-space.

13. Lucy Williamson, "North Korea Defies Warnings in Rocket Launch Success," BBC News, December 12, 2012, https://www.bbc.com/news/world-asia-20690338.

14. John Schilling, "North Korea's Space Launch: An Initial Assessment," 38 North, February 9, 2016, https://www.38north.org/2016/02/jschilling020816.

15. Eric Talmadge, "North Korea Hopes to Plant Flag on the Moon," Associated Press, August 4, 2016, https://web.archive.org/web/20160820032940/http://www.msn.com/en-us/news/world/ap-exclusive-north-korea-hopes-to-plant-flag-on-the-moon/ar-BBvePu4?li=BBnbcA1.

16. Adelle Nazarian, "Why the 'Failed' North Korean Missile Launch Can Still Be a Threat," Observer Research Foundation, June 1, 2017, https://www.orfonline.org/expert-speak/failed-north-korean-missile-launch-can-still-threat. Brackets in original.

17. Defense Intelligence Agency, "Challenges to Security in Space" (Washington DC: Defense Intelligence Agency, January 2019), https://www.dia.mil/Portals/27/Documents/News/Military%20Power%20Publications/Space_Threat_V14_020119_sm.pdf.

18. Defense Intelligence Agency, "Challenges to Security," iii.

19. Wood, "Countries with the Most."

20. Defense Intelligence Agency, "Challenges to Security."

21. Zachary Keck, "China Will Soon Be Able to Destroy Every Satellite in Space and the U.S. Military Should Be Worried," *National Interest*, July 30, 2018, https://nationalinterest.org/blog/buzz/china-will-soon-be-able-destroy-every-satellite-space-27182.

22. Michael Sheetz and Amanda Macias, "China and Russia Are Militarizing Space with 'Energy Weapons' and Anti-Satellite Missiles," CNBC, February 13, 2019, https://www.cnbc.com/2019/02/13/pentagon-warns-of-weaponization-of-space-by-china-russia-report.html.

23. Sheetz and Macias, "China and Russia."

24. "Do All Satellites Have to Fly at the Same Speed so Not to Leave Their Orbit?," How Things Fly, Smithsonian National Air and Space Museum, January 17, 2013, https://

howthingsfly.si.edu/ask-an-explainer/do-all-satellites-have-fly-same-speed-so-not-leave
-their-orbit.

25. Kasey Cordell, "How Air Force Space Aggressors Protect Our Far-Flung Assets—
and Way of Life," *5280*, May 2018, https://www.5280.com/2018/04/how-the-air-forces
-space-aggressors-are-helping-protect-our-far-flung-assets-and-way-of-life.

26. Commission to Assess United States National Security Space Management and
Organization, "Report of the Commission to Assess United States National Security Space
Management and Organization" (Washington DC, January 11, 2001), https://aerospace.csis
.org/wp-content/uploads/2018/09/RumsfeldCommission.pdf.

27. Craig Covault, "Chinese Test Anti-Satellite Weapon," *Aviation Week & Space
Technology*, January 17, 2007, https://web.archive.org/web/20070128075259/http://
www.aviationweek.com/aw/generic/story_channel.jsp?channel=space&id=news
%2FCHI01177.xml.

28. "UCS Satellite Database: In-Depth Details on the 2,666 Satellites Currently Orbiting Earth, Including Their Country of Origin, Purpose, and Other Operational Details,"
Union of Concerned Scientists, December 8, 2005, updated April 1, 2020, https://www
.ucsusa.org/resources/satellite-database; and "Iran Launches Its First Military Satellite," *Al
Jazeera*, April 22, 2020, https://www.aljazeera.com/news/2020/04/iran-launches-military
-satellite-200422062126491.html.

29. Mildred Sola Neely and Kathleen J. Brahney, "U.S. Laser Weapon Test: 'Star Wars
Alive and Well'?," Federation of American Scientists, October 22, 1997, https://fas.org/spp
/military/program/asat/971022-miracl-mr.htm; and Gerry Doyle, "Anti-satellite Weapons: Rare, High-Tech, and Risky to Test," Reuters, March 27, 2019, https://www.reuters
.com/article/uk-india-satellite-tests-factbox/anti-satellite-weapons-rare-high-tech-and
-risky-to-test-idUKKCN1R80Q1.

30. Doyle, "Anti-satellite Weapons."

31. Gideon Brough and Alun Williams, "Decapitation Strategy," in *The Encyclopedia of
War*, ed. Gordon Martel (Oxford: Blackwell Publishing, John Wiley, November 13, 2011),
https://onlinelibrary.wiley.com/doi/abs/10.1002/9781444338232.wbeow163.

32. Commission to Assess United States National Security Space Management and
Organization, "Report."

33. David Axe, "NATO Fears Russia Jamming Its GPS Systems in a War: They Might
Have a Solution," *National Interest*, June 7, 2019, https://nationalinterest.org/blog/buzz
/nato-fears-russia-jamming-its-gps-systems-war-they-might-have-solution-61437.

34. Patrick Tucker, "Pentagon Shelves Neutral Particle Beam Research," *Defense One*,
September 4, 2019, https://www.defenseone.com/technology/2019/09/pentagon-shelves
-neutral-particle-beam-research/159643.

35. "X-37B Orbital Test Vehicle," U.S. Air Force, September 1, 2018, https://www.af.mil
/About-Us/Fact-Sheets/Display/Article/104539/x-37b-orbital-test-vehicle.

36. "X-37B Orbital Test Vehicle."

37. Jackie Wattles, "Air Force's Mysterious X-37B Spacecraft Sets New Record for Time
in Space," CNN Business, August 28, 2019, https://www.cnn.com/2019/08/27/tech/x-37b
-air-force-space-plane-days-in-space-scn-trnd/index.html.

38. Valerie Insinna, "The U.S. Air Force's x-37b Spaceplane Lands after Spending Two Years in Space," *Defense News*, October 28, 2019, https://www.defensenews.com/space/2019/10/28/the-air-forces-x-37b-spaceplane-finally-landed-after-spending-two-years-in-space/.

39. Sebastien Roblin, "Russia's 'Killer' Space Satellites: A Real Threat or a Paper Tiger?," *National Interest*, August 24, 2018, https://nationalinterest.org/blog/buzz/russias-killer-space-satellites-real-threat-or-paper-tiger-29717.

40. David Axe, "Russia Just Tested a Killer Satellite," *National Interest*, April 16, 2020, https://nationalinterest.org/blog/buzz/russia-just-tested-satellite-killer-144797.

41. Defense Intelligence Agency, "Challenges to Security."

42. Sandra Erwin, "Trump Signs Defense Bill Establishing U.S. Space Force: What Comes Next," *Space News*, December 20, 2019, https://spacenews.com/trump-signs-defense-bill-establishing-u-s-space-force-what-comes-next/.

43. Oriana Pawlyk, "It's Official: Trump Announces Space Force as 6th Military Branch," Military.com, June 18, 2018, https://www.military.com/daily-news/2018/06/18/its-official-trump-announces-space-force-6th-military-branch.html.

44. Nayef R. F. Al-Rodhan, "U.S. Space Policy and Strategic Culture," *Columbia Journal of International Affairs*, April 16, 2018, https://jia.sipa.columbia.edu/online-articles/us-space-policy-and-strategic-culture.

45. "Thucydides's Trap," Harvard Kennedy School, Belfer Center for Science and International Affairs, https://www.belfercenter.org/thucydides-trap/overview-thucydides-trap (accessed October 9, 2019).

46. Graham Allison, "The Thucydides Trap," *Foreign Policy*, June 9, 2017, https://foreignpolicy.com/2017/06/09/the-thucydides-trap. See also Graham Allison, *Destined for War: Can America and China Escape Thucydides's Trap?* (Cambridge: Belfer Center for Science and International Affairs, 2017).

47. Hans M. Kristensen and Matt Korda, "Status of World Nuclear Forces," Federation of American Scientists, April 2020, https://fas.org/issues/nuclear-weapons/status-world-nuclear-forces/.

12. The Fate of Humanity

1. Sokolski, *Getting Mad*.

2. Laura Geggel, "How Do Intercontinental Ballistic Missiles Work?," Live Science, November 30, 2017, https://www.livescience.com/61062-how-do-intercontinental-ballistic-missiles-work.html.

3. Geggel, "How Do Intercontinental Ballistic Missiles Work?"

4. Tim Kelly, "The MAD Myth," Future of Freedom Foundation, August 10, 2012, https://www.fff.org/explore-freedom/article/the-mad-myth.

5. Geggel, "How Do Intercontinental Ballistic Missiles Work?"

6. Neely and Brahney, "U.S. Laser Weapon Test."

7. Henry T. Nash, *Nuclear Weapons and International Behavior* (Leyden, Netherlands: A. W. Sijthoff International Publishing, 1975).

8. "Treaty between the United States of America and the Union of Soviet Socialist Republics on the Limitation of Anti-Ballistic Missile Systems," U.S. Department of State,

Bureau of International Security and Nonproliferation, May 26, 1972, https://2009-2017
.state.gov/t/isn/trty/16332.htm.

9. "Fact Sheet: Memorandum of Understanding on Succession," U.S. Department of
State, September 26, 1997, https://1997-2001.state.gov/www/global/arms/factsheets
/missdef/mou.html.

10. "ABM Treaty Fact Sheet: Announcement of Withdrawal from the ABM Treaty," Office
of the Press Secretary, White House, press release, December 13, 2001, https://georgewbush
-whitehouse.archives.gov/news/releases/2001/12/20011213-2.html.

11. Dave Majumdar, "Russia's Nuclear Weapons Buildup Is Aimed at Beating U.S. Missile Defenses," *National Interest*, March 1, 2018, https://nationalinterest.org/blog/the-buzz
/russias-nuclear-weapons-buildup-aimed-beating-us-missile-24716.

12. Neely and Brahney, "U.S. Laser Weapon Test."

13. David Axe, "The Pentagon Has a Plan to Arm Satellites with Lasers to Shoot Down
Missiles; It's Insane," *Daily Beast*, August 20, 2018, https://www.thedailybeast.com/the
-pentagon-has-a-plan-to-arm-satellites-with-lasers-to-shoot-down-missiles-its-insane.

14. "The Biological Weapons Convention: Convention on the Prohibition of the Development, Production and Stockpiling of Bacteriological (Biological) and Toxin Weapons
and on Their Destruction," United Nations, https://www.un.org/disarmament/wmd/bio
(accessed October 16, 2019).

15. "We Can Eliminate Nuclear Weapons in Our Lifetime," Global Zero, https://www
.globalzero.org (accessed October 16, 2019).

16. Pellerin, "Deputy Secretary."

17. "Thoughts on the Business of Life: Calvin Coolidge," Forbes Quotes, https://www
.forbes.com/quotes/1385 (accessed October 27, 2019).

Appendix A

1. Malcolm Scott and Cedric Sam, "Here's How Fast China's Economy Is Catching up
to the U.S.," Bloomberg, May 12, 2016, updated May 21, 2019, https://www.bloomberg.com
/graphics/2016-us-vs-china-economy.

2. Sydney J. Freedberg Jr., "U.S. Defense Budget Not That Much Bigger than China,
Russia: Gen. Milley," *Breaking Defense*, May 22, 2018, https://breakingdefense.com/2018
/05/us-defense-budget-not-that-much-bigger-than-china-russia-gen-milley.

3. Freedberg, "U.S. Defense Budget."

4. Freedberg, "U.S. Defense Budget."

5. Freedberg, "U.S. Defense Budget."

Appendix B

1. A. E. Siegman, *Lasers* (Sausalito CA: University Science Books, 1986), 2.

2. "How Lasers Work," Lawrence Livermore National Laboratory, National Ignition
Facility & Photon Science, https://lasers.llnl.gov/education/how_lasers_work (accessed
June 26, 2019).

INDEX

Page numbers in italics refer to illustrations.

ABM Treaty (1972). *See* Anti-Ballistic Missile Treaty (1972)

accelerating returns, law of, 74–75, *76*

Active Denial System, 96, 106

active protection system (APS), 171–72

Acton, James, 25

advertising, 98–99

Aegis combat system, 79–81, 143–44

Afganit APS, 171–72

Afghanistan, 96, 160, 168

AI (artificial intelligence): dangers of, 9, 184–85, 213; in dual-use technology, 57–58; ethics and, 182–83; human intelligence compared to, 180–82; in malware, 148–49; military use of, 25, 179–80, 182–83, 217, 219; pace of warfare and, 2, 8; smart agents and, 148; Third U.S. Offset Strategy and, 2–3, 55–56, 58, 219

aircraft carriers, 79, 80, 143, 184, 204

algorithms, computer, 148, 180, 218, 219, 220. *See also* AI (artificial intelligence)

alliances, 59–61

Allison, Graham, 211

All-Union Conference on Radio-Spectroscopy (1952), 66

al-Qaeda, 160

altitude, effects of, 116–17, 119, 124–25, 156, 162, 164

American Institute of Physics, 68

American Revolution, 35–36

antiballistic missile systems, 6, 20, 73, 86, 216–17

Anti-Ballistic Missile Treaty (1972), 20, 216–17

APS. *See* active protection system (APS)

Arena (active protection system upgrade), 171

armored vehicles, 162–63, 168–69, 172

Army Technology (website), 139

artificial intelligence. *See* AI (artificial intelligence)

The Artificial Intelligence Revolution (Del Monte), 145

The Art of War (Sun Tzu), 5, 35, 153, 179

Asia-Pacific region, 36, 84–85, 149, 165

Assad, Bashar, 19

atomic bombs, 13–14, 19, 29–30, 48, 116–17

Atomic Heritage Foundation, 73

atoms, 29, 68, 124–25, 169

aurora borealis, artificial, 118

autonomous weapon systems. *See* AWS (autonomous weapon systems); weapons, autonomous

Aviation Week, 201–2

AWS (autonomous weapon systems), 180. *See also* weapons, autonomous

B83 (bomb), 30

balance of power, 18–19, 211–12

balance of terror, 10, 15, 18–19, 26, 86, 214

Basov, Nikolay, 66–67, 69

Battle of Britain, 139–41

Belarus, 136, 216

Bell, Alexander Graham, 50

Bell Laboratories, 68–69

Biological Weapons Convention, 219

Bitcoin, 132–33

Bizarre Project, 90

"black hats" (hackers), 137, 158

black programs, 45, 90

blindness in humans, 71–72, 74

Blitzer, Wolf, 129

Boeing, 100–101, 168–70, 171, 175

bombs, 2, 13–14, 19, 29–30, 48, 116–17, 179–80

Booz Allen Hamilton (consulting firm), 78

Bostrom, Nick, 181

brain damage, 90, 92, 94, 95, 97–98

Breaking Defense, 81, 224

Brennan, John, 145

Britain. *See* Great Britain

Brown, Harold, 52

bug bounties, 158, 159

Bulletin of the Atomic Scientists, 28

Bush, George W., 20, 201, 216

Business Insider, 142

c (speed of light), 8, 14. *See also* light, speed of

Capozzella, Robert J., 105

Carter, Ash, 53, 55

Carter, Jimmy, 52

C-CLAW (close combat laser assault weapon), 71–72

CCTV (China Central Television), 83

Central Intelligence Agency. *See* CIA (Central Intelligence Agency)

CERN (European Organization for Nuclear Research), 175

Cesaro, Richard, 90

CHAMP (Counter-Electronics High Power Microwave Advanced Missile Project), 100–101, 106

China: AI and, 181; American-based businesses of, 57–58; asymmetrical warfare approach of, 5–6, 22–23; automation and, 25–26; autonomous weapons and, 221; countermeasures by, 154–56; cyberspace abilities of, 21, 128, 138, 145–48, 149–50; defense budget of, 41, 223–24; election interference by, 82; EMP weapons of, 112; espionage methods of, 145–47; Korean War and, 49; laser weapons and, 82–84, 88, 194, 218; MAD and, 20–21; microwave weapons of, 92, 104, 106–7; missiles of, 24, 36, 79, 194; nuclear war effects on, 32; nuclear weapons of, 28, 31, 51; pace of war and, 34; rail guns and, 24; Russia and, 104, 107; satellites of, 209; space assets of, 202–3; space weap-

onization by, 199–202, 210; treaties and, 72, 197–98; United States and, 84–85, 150, 165–66, 211–12, 223–24; weapons buildup of, 4; as world power, 19, 58–61

China Sea, 28, 59, 84, 85

Churchill, Winston, 139

CIA (Central Intelligence Agency), 130, 158

Clinton, Bill, 56–57

Clinton, Hillary, 82, 129–32, 134

close combat laser assault weapon. *See* C-CLAW (close combat laser assault weapon)

CNBC, 36, 99, 200

CNN, 2, 101, 129, 207

Cold War: balance of terror during, 18; human judgment during, 8–9, 215; Korea and, 48–49; MAD and, 79, 214; SDI and, 73; weapons and, 51–52, 89, 91, 119, 120; World War II remembered during, 27

Cold War II, 19–20, 58–59

Coleman, Keith, 101

combat systems, 79–81, 143–44

Combined Space Operations Center. *See* CSPOC (Combined Space Operations Center)

Comey, James, 130

computers, 74, 135–37, 148, 150, 181

Congress, 125, 144, 217

Conventional Cold War, 58–59. *See also* Cold War II

Convention on Prohibitions or Restrictions on the Use of Certain Conventional Weapons (1980), 72

Coolidge, Calvin, 221

Cooper, Robert S., 54–55

Council on Foreign Relations, 130

Counter-Electronics High Power Microwave Advanced Missile Project. *See* CHAMP (Counter-Electronics High Power Microwave Advanced Missile Project)

Crimea annexation, 142

cryptocurrency, 132–33

CSPOC (Combined Space Operations Center), 194

CTV, 15

Cuba, 91–92, 104, 119–20

Cuban Missile Crisis, 119–20

c-war (war at the speed of light), 8, 213, 218, 219–22

Cyber Command, 138
cyberspace as battlefield, 7–8, 21, 121, 138

Daily Beast, 217
Daily Mail, 100–101
DARPA (Defense Advanced Research Projects
    Agency), 25, 53–55, 56, 57, 89–90, 95
dazzlers (low-power lasers), 4, 43, 74, 76–77
debris fields, 200–201, 202, 203, 204, 210, 212
Defense Advanced Research Projects Agency.
    See DARPA (Defense Advanced Research
    Projects Agency)
Defense Digital Service, 159
Defense Innovation Unit. *See* DIU (Defense
    Innovation Unit)
Defense Intelligence Agency, 84, 199
Defense News, 53–54
defense systems, 105, 125, 172, 189, 190
Del Monte, Louis A., 76, 98, 145
Del Monte & Associates, Inc., 98
Democratic National Committee (DNC), 82,
    134, 135
Democratic Party, 130–31, 134–35
Department of Defense (DOD): AI and, 179;
    autonomous weapons and, 149, 180, 183–
    86; budget of, 40–41, 56–57, 159–60, 223–24;
    China and, 147–48; cyberspace warfare and,
    138; directed-energy weapons and, 7–8, 38,
    71, 80, 157; electronic warfare and, 144, 149;
    military-industrial complex and, 40–41, 160–
    61, 171; politics influencing, 56–57; priorities
    of, 191, 217; Second U.S. Offset Strategy and,
    52; semiautonomous weapons and, 183–86;
    Soviet Union and, 85; Third U.S. Offset Strat-
    egy and, 2. *See also* Pentagon
Department of Homeland Security (DHS), 128,
    133, 159
Department of Justice, 146
DEWS. *See* weapons, directed-energy
DHS. *See* Department of Homeland Security (DHS)
Diana, HMS, 139
Diplomat, 23, 52–53
Directive 3000.09, 149, 183–86, 187
Discover, 118
DIU (Defense Innovation Unit), 56, 57
DNC. *See* Democratic National Committee (DNC)
DOD. *See* Department of Defense (DOD)
Dong Feng-26 (missile), 5–6

Dong Mingzhu, 181
doomsday clock, 28
drones, 6–7, 43, 101–3, 162, 184–85, 210
Drozd (active protection system), 171
dual-use technology, 40, 57–58, 196, 197, 209
Dulles, John Foster, 49–50
Dunford, Joseph F., 35

EA-18G Growler aircraft, 162
Earth, human life on, 21, 32, 118, 221
Earth, surroundings of, 174, 175
Economist, 6, 194
Edison, Thomas, 114
Einstein, Albert, 8, 13–14, 36, 66, 67
Eisenhower, Dwight, 38–39, 48, 49–50
elections, presidential, 82, 129–32, 133–35
electric fields, 18, 143
electromagnetic fields, 7, 23, 156, 157, 205
electromagnetic pinch devices, 122
electromagnetic pulse. *See* EMP (electromag-
    netic pulse); weapons, EMP
electromagnetic spectrum, 7, 141, 142–44, 158,
    164, 173
electronics, 67, 107–9, 156, 163, 195
Electron Tube Research Conference (1952), 66
Elta Systems, 172
emails, 130, 131–32
embassies and embassy personnel, 89, 91–92,
    104, 107
EMP (electromagnetic pulse), 31–32, 108, 112–
    16, 127, 198. *See also* weapons, EMP
Encyclopaedia Britannica, 122
Enterprise (fictional starship), 1, 14–15, 167
espionage, 37, 50, 58, 83, 104, 134, 145–47
ethics, 2, 72, 182–83, 185–86, 188–89
European Organization for Nuclear Research.
    See CERN (European Organization for
    Nuclear Research)
European Union, 19
Executive Order on Coordinating National Resil-
    ience to Electromagnetic Pulses, 127–28, 158

F-35 (aircraft), 39
false alarms, 8–9, 214–15
famine, 31, 32, 33–34
Faraday, Michael, 114
Faraday cages, 18, 108, 156–57
Fat Man (atomic bomb), 29–30

FBI (Federal Bureau of Investigation), 130, 134–35
Fedoseyev, Anatoly, 86
feedback element, 225, 226
Fermi, Enrico, 116–17
Financial Brand, 98
Finlay, Peter, 101
First Gulf War, 2, 52, 73–74, 157, 214
First U.S. Offset Strategy, 48–50, 51–52
Fisher, Richard D., Jr., 106
FiveThirtyEight, 130
5280, 201
Flash Gordon, 75, 76
flux compression generators, 121
force fields: future of, 175–76; matter and, 168–72; predicted, 14–15, 167–68; radiation and, 172–75
Forstchen, William R., 124
Fourth U.S. Offset Strategy, 59–61, 219
France, 31, 51, 139
Frey, Allan H., and Frey effect, 92, 96
Future of Life (organization), 182
Futurism (website), 105

gamma radiation, 18, 124–25
Geneva Conference on Disarmament (2018), 209
Geneva Convention (1977), 218
Georgia (nation), 59, 142
Germany, 139–41
glasnost, 85–86
Global Hawk, 210
Global Positioning System. *See* GPS (Global Positioning System)
Global Times, 106
Gordon, James, 67
Gould, Gordon, 68–69, 71
GPS (Global Positioning System), 142, 143, 148, 196, 197, 201, 205–6
Great Britain, 139–41
Griffin, Michael, 190, 206
Groves, L. R., Jr., 38

hacking, 82, 83, 104–5, 133–35, 137, 158–59, 205
Hack the Department of Homeland Security Act (2018), 159
Hagel, Chuck, 2–3, 52–53
Harris, Harry, 36
Hashemipour, Seyed, 77–78
Havana Syndrome, 91–92, 94, 95, 97, 104
Hawaii, 117–18, 123

Hawkins, Adrian, 134
HEL (high-energy laser) weapons, 78
Helena, USS, 49
Hero of Socialist Labor Award, 86
hertz as energy force measurement, 92–93
Hill, 28, 59–60
Hiroshima, Japan, 13, 28–29, 30
History Learning Site, 140
Hitler, Adolf, 139–41
Honeywell, 39–41, 54–55, 57, 98, 108, 160–61, 195
HRP (Human Research Program), 174
Huang Chenguang, 155
Hughes, Emmet J., 49
Hughes Aircraft, 161
Hughes Laboratories, 69–70
human intelligence, 2, 3, 180–82
human judgment, 8–9, 149, 181–82, 183–84, 211, 214–15, 218–19, 221
Human Research Program. *See* HRP (Human Research Program)
hydrogen bombs. *See* weapons, thermonuclear

ICBMs (intercontinental ballistic missiles), 5, 6, 22, 51, 79, 200, 215–16
IEDs (improvised explosive devices), 168
IMI. *See* Israel Military Industries (IMI)
improvised explosive devices (IEDs), 168
India, 31, 194
Indian Journal of Biochemistry & Biophysics, 90
integrated circuits, 74, 75, 108, 157
intelligent agents, 148, 180–81, 220. *See also* AI (artificial intelligence)
intercontinental ballistic missiles. *See* ICBMs (intercontinental ballistic missiles)
International Security, 20–21
International Space Station, 203
International Traffic in Arms Regulations (ITAR) form, 39–40
Internet Research Agency, 133
Invention Secrecy Act (1951), 97
Iran: CHAMP and, 100; Fourth U.S. Offset Strategy and, 79; MAD and, 216; satellites of, 202–3; Stuxnet and, 136–37; swarming by, 101–2, 186; terrorist organizations and, 82; weapons of, 112, 126–27
Iraq, 2, 74, 157, 160, 168, 214
Iraq War, 168, 214
Iron Fish (missile interceptor system), 172

Islamic State, 138
Israel, 31, 51, 137–38, 172
Israel Defense, 172
Israel Military Industries (IMI), 172

jamming, 101, 142, 143–44, 160, 162–64, 165,
    195, 205–6
Jane's 360, 82–83
Japan, 13, 28–29, 30
Joint Comprehensive Plan of Action
    (JCPOA), 126
joules as energy force measurement, 29–30

Kazakhstan, 85, 119, 216
Kennedy, John F., 120
Kessler, Ronald, 100
Khrushchev, Nikita, 120
Kim Il Sung, 49
kinetic energy, 14, 23, 200
Klunder, Matt, 23
KMS satellites, 126
Korean War, 47, 49–50
Koslov, Sam, 90
K Project, 119, 120
Kurzweil, Ray, 74–75

lasers, 65–70, 225–26. *See also* weapons, laser
laser weapon system. *See* LAWS (laser weapon
    system)
laser winter, 59
Lawrence Livermore National Laboratory
    (LLNL), 37, 42–43, 73
LAWS (laser weapon system), 75–76, 78, 83
LAWS (lethal autonomous weapons systems),
    182–83
Lenin Prize, 86
LEO. *See* low-Earth orbit (LEO)
lethal autonomous weapons systems. *See* LAWS
    (lethal autonomous weapons systems)
Lexico, 172–73
light, speed of: autonomous weapons and, 187;
    cyber weapons traveling at, 138, 185; directed-
    energy weapons traveling at, 4; lasers trav-
    eling at, 77, 194, 204, 217; laser weapons
    traveling at, 186; in mass to energy conver-
    sion, 14; microwaves traveling at, 89, 204; war
    at, 8, 34, 42
lightning, 112, 114, 116–17
Little Boy (atomic bomb), 29–30

LLNL. *See* Lawrence Livermore National Labo-
    ratory (LLNL)
Lockheed Martin, 4, 39, 80, 81, 155, 160, 161–
    62, 189, 217
Long Peace, 27
Los Angeles Times, 132
low-Earth orbit (LEO), 21, 118, 119, 121, 194, 198,
    200, 202, 207
Lukashev, Aleksey, 131–32
Lukasik, Stephen, 90

machine learning, 181
MAD (mutually assured destruction), 1, 10, 20–
    22, 24–26, 79, 101, 126–27, 213–18
magnetic fields, 32, 39–40, 114–15, 116, 121–22,
    124, 143, 173, 174, 175–76
malware, 132, 136–38, 148–49
Manhattan Project, 37–38, 73
Maritime Executive, 83
Mark 50 (torpedo), 45
Marx, Erwin Otto, 122
Marx generators, 122
masers, 66–68
mass-energy equivalent formula, 8, 13–14
Medium, 168
MEDUSA (Mob Excess Deterrent Using Silent
    Audio), 96–97
Microsoft, 98, 158, 159
microwaves, 89, 92–94. *See also* weapons,
    microwave
military, U.S.: AI and, 3, 9; autonomous weap-
    ons and, 184–85, 187, 188–89, 219–21; China
    as threat to, 84, 202; countermeasures by, 6,
    107–9, 157; cyberspace and, 7, 128; electronic
    warfare and, 141, 143–44, 149–50, 160; EMP
    weapons and, 117–19; laser weapons and, 71–
    72, 73–74, 75, 77–78, 156, 190, 194, 217–18;
    MAD and, 19–20; microwave weapons and,
    96, 101–2, 104–5, 194; missile attacks and,
    202–5; nanoweapons and, 76; nuclear weap-
    ons and, 29–30; politics influencing, 56; satel-
    lites of, 196–97; secrecy and, 37–38, 47, 104–5;
    space assets of, 199, 210; speed of warfare and,
    34–37; strategies of, 22–24, 47–48, 210–11. *See
    also* Department of Defense (DOD); military-
    industrial complex; U.S. Air Force; U.S.
    Army; U.S. Navy
Military & Aerospace Electronics, 105

military-industrial complex, 38–41, 48, 56, 160–61, 188

Military Review, 186

Military Strategic and Tactical Relay system. *See* Milstar (Military Strategic and Tactical Relay) system

Millennium Challenge (war game), 102

Miller, Stephanie, 96

Mills, Michael J., 33–34

Milstar (Military Strategic and Tactical Relay) system, 44–45, 193–96

mirrors, 37, 42, 225–26

Missile Defense Agency, 216

missiles, anti-satellite. *See* weapons, anti-satellite

missiles, ballistic, 5, 6, 22, 51, 79, 119–20, 125, 126, 200, 215–17

missiles, cruise, 78, 81, 82, 101, 188, 189

missiles, hypersonic, 4, 6, 24–25, 36–37, 43, 190–91, 217

missiles, intercontinental ballistic. *See* ICBMS (intercontinental ballistic missiles)

Missile Threat, 5–6

Mob Excess Deterrent Using Silent Audio. *See* MEDUSA (Mob Excess Deterrent Using Silent Audio)

Mohammadian, N., 77–78

Moore, Gordon, and Moore's law, 74–75

Moscow Signal, 89–90, 91, 92, 95, 97, 104

Mueller, Robert, 131, 133, 138

Müller, Vincent, 181

Nagasaki, Japan, 13, 28–29, 30

nanotechnology and nanoweapons, 76

Nanoweapons (Del Monte), 76, 145

NASA (National Aeronautics and Space Administration), 15, 174, 207

Natanz nuclear facility, 136

National Aeronautics and Space Administration. *See* NASA (National Aeronautics and Space Administration)

National Climate Assessment, 115–16

National Defense Authorization Act (NDAA), 80, 144, 210

National Defense Strategy (2018), 60–61, 144

National Electrical Manufacturers Association Surge Protection Institute, 114

National Ignition Facility. *See* NIF (National Ignition Facility)

National Interest, 28, 189, 200, 209

National Nanotechnology Initiative. *See* NNI (National Nanotechnology Initiative)

National Public Radio. *See* NPR (National Public Radio)

National Security Agency, 130, 147

National World War II Museum, 48

NATO (North Atlantic Treaty Organization), 7, 19, 23, 52, 79, 138, 142, 157, 205–6

Nazis, 139–41

NBC, 15, 92

NDAA. *See* National Defense Authorization Act (NDAA)

New Look strategy, 50. *See also* First U.S. Offset Strategy

New Scientist, 96–97

Newsweek, 87, 126

New York Times: on China, 84; on cyberattacks, 147, 159; on Havana Syndrome, 91, 92; on Millennium Challenge, 102; on Russian hacking, 82, 133, 134–35; on Soviet laser weapons, 85–86; on Stuxnet, 137; on temperature-related deaths, 112–13, 115–16

Next Big Future (website), 190

NIF (National Ignition Facility), 42–43

Nixon, Richard, 127

NNI (National Nanotechnology Initiative), 76

Nobel Prize, 67–69, 121

Non-Proliferation Treaty, 216

NORAD (North American Aerospace Defense Command), 215

North Atlantic Treaty Organization. *See* NATO (North Atlantic Treaty Organization)

North Korea: CHAMP and, 100–101; Cold War and, 48–49; Fourth U.S. Offset Strategy and, 60; MAD and, 214, 216; satellites of, 197–98, 202–3; South Korea and, 50; weapons of, 31, 78–79, 112, 121, 125–27, 128, 198

Northrop Grumman, 160, 184–85

NPR (National Public Radio), 57–58

nuclear fission, 13, 29, 30, 51

nuclear fusion, 30, 42–43, 51, 122

nuclear testing, 117–20

nuclear war: barely avoided, 8–10, 119–20, 214–15; EMP weapons and, 125–26; history of, 28–

29; MAD and, 1, 22, 25–26, 79, 213–14, 215–16, 217–18; present possibility of, 215–16; results of, 31–34, 211–12

nuclear winter, 31–34, 59, 203, 212, 214, 219

Obama, Barack, 53–54, 56, 57, 91

Observer Research Foundation, 198

Office of Naval Research (ONR), 25

Office of the Director of National Intelligence, 130, 134

offset strategies, 2–3, 25, 47–53, 55–58, 59–61, 216

One Second After (Forstchen), 124

ONR. *See* Office of Naval Research (ONR)

Operation Burnt Frost, 203

Operation Desert Storm (1991), 179–80. *See also* First Gulf War

orbit, geostationary, 193–95

orbit, geosynchronous, 194

orbit, low-Earth. *See* low-Earth orbit (LEO)

Orbital Test Vehicle. *See* X-37B (Orbital Test Vehicle)

Ørsted, Hans Christian, 143

Outer Space Treaty, 22, 197–98, 199–201

ovens, microwave, 93–94

Pakistan, 31

Pandora Project, 90, 95

Partial Test Ban Treaty (1963), 120

particle beam weapons, 73, 206

Patent and Trademark Office, 51, 69, 97, 170

patents, 50–51, 68–69, 97, 114, 168–71, 175

Patriot missiles, 6, 186

Peace Pledge Union, 19

Pearl Harbor attack, 13, 47–48, 202

Pearson, Lester, 18–19

Pentagon, 36, 52, 53, 78–79, 83, 90, 210. *See also* Department of Defense (DOD)

People's Liberation Army (PLA), 147, 154

People's Liberation Army Navy (PLAN), 82–83

Peresvet (laser weapon), 87

Petrov, Stanislav, 8–9, 214

Phalanx weapon system, 103, 182, 186–87

phishing, 131–32

Phys.org, 36, 96

pinch effect, 122

PLA. *See* People's Liberation Army (PLA)

PLAN. *See* People's Liberation Army Navy (PLAN)

plasma, 42, 75, 122, 155, 169–71, 174, 175, 204

Poblete, Yleem, 209

Podesta, John, 131–32

Politico, 135

Ponce, USS, 4, 22, 75–76, 77, 78, 155

Popular Science, 106

power losses, 112–14, 115–16

power surges, 114–15, 156–57

PPWT treaty, 22

Preble, USS, 4

Prigozhin, Yevgeny, 133

Prokhorov, Alexander, 66–67, 69

Protocol on Blinding Laser Weapons, 72

Pry, Peter Vincent, 32, 198

Putin, Vladimir, 24–25, 36, 87–88, 133–34, 142, 216

Quartz (news organization), 162

radar, 93, 140–41, 164–65, 171–72

radiation, 17–18, 29, 107–9, 118, 124–25, 157, 172–75, 195

radiation-hardened electronics, 107–9, 156, 160, 195

radio waves, 73, 89, 92–93, 101, 140, 142–43, 173

Rafael Advanced Defense Systems, 172

rail guns, 23–24

Ranets-E (defense system), 105

Raymond, John, 209

Raytheon, 102, 160, 161

RCA, 67

Reagan, Ronald, 52, 73

Red Cross, 183

Republican Party, 134

Reuters, 36, 101–2

RFPS (requests for proposals), 160–61

Rhee, Syngman, 49

Robock, Alan, 32

robots, killer. *See* LAWS (lethal autonomous weapons systems)

Roddenberry, Gene, 1, 15, 167–68

Roosevelt, Franklin, 48

Roper, William, 55

Rosa-E (defense system), 105

Rubáiyát of Omar Khayyám, 77

Rumsfeld, Donald, 201

Russia: active protective systems of, 171–72; asymmetrical warfare approach of, 5–6, 22–23; automation and, 25–26; autonomous weapons and, 188, 221; China and, 92, 104, 107; countermeasures against, 159–60, 161–62; Cuba and, 92, 104; cyberspace abilities of, 21, 128, 138, 141–42, 143–44, 149–50, 205–6; defense systems of, 189; election interference by, 82, 130–32, 133–35; EMP weapons of, 112, 125; laser weapons of, 82, 85–88, 194; MAD and, 20–21; microwave weapons of, 92, 104–5, 107; missiles of, 24–25, 36; nuclear war effects on, 31–32; nuclear weapons of, 27–28, 31; pace of war and, 34; population of, 31; satellites of, 209; space assets of, 202–3; space weaponization by, 199–201, 210; treaties and, 72, 197, 216; Ukraine and, 160; weapons buildup of, 4, 79, 218; as world power, 19–20, 58–61. *See also* Soviet Union

Russo-Georgian War (2008), 142

Russo-Japanese War (1904–5), 139

S-400 (defense system), 189, 190

S-500 (defense system), 189

Saber Fury, 162–63

Sagan, Carl, 41, 142

Sakharov, Andrei, 121

Salmanogli, Ahmad, 77–78

Sandia National Laboratories, 122

satellites, 21, 44–45, 118, 126, 143, 193–98, 200–205, 207, 209. *See also* weapons, anti-satellite

Schawlow, Arthur, 68–69

Schultz, Debbie Wasserman, 135

Science Alert (website), 175

Science Clarified (website), 71

science fiction as predictor, 1, 15, 37, 45, 76–77, 167–68, 175

SCO (Strategic Capabilities Office), 53–54, 55–56, 57

SDI (Strategic Defense Initiative), 73

Sea Hunter (ship), 25

Second Gulf War. *See* Iraq War

Second U.S. Offset Strategy, 52–53

Secret Service, 97

Selva, Paul, 144

Senate, 82, 159, 190

sensors, 39–40, 54, 170–71, 196

shielding, 107, 108, 117, 169–71, 172, 174–76

shock waves, 168–69, 171, 172

Shoigu, Sergei, 24

Sierra Nevada Corporation, 96

signals intelligence, 139, 163

Silicon Valley, 56

Sina.com, 83

skin irritation, 96, 106

Slickdeals.net, 99

smart agents. *See* AI (artificial intelligence); intelligent agents

smart bombs, 2, 179–80

Smithsonian, 120

snowstorms, 112–14

SOI CMOS technology, 195

solar flares, 114, 127, 173, 174

sound, speed of, 4, 23, 24, 34, 191, 207, 209

sounds, mysterious, 91–92, 96–97, 98

South China Morning Post, 154–55

South Korea, 48–49, 50

Soviet Union: China compared to, 58; Cold War and, 8–9, 18–19, 79, 214–15; collapse of, 27, 41, 52, 160; Cuban Missile Crisis and, 119–20; in Korean War, 49; microwave use by, 89, 90–91; nuclear testing and, 118, 119–20; research in, 66–67; on SDI, 73; treaties and, 20, 216; weapons of, 37, 48, 50, 51–52, 85–86, 106, 121. *See also* Russia

space assets, 22, 149, 194–96, 199–200, 201, 202–6, 210–11, 212

space law, 197, 209

Space Radiation Superconducting Shield. *See* SR2S (Space Radiation Superconducting Shield)

spear phishing, 131–32

Spencer, Richard, 83

spoofing, 164, 201, 205–6

SR2S (Space Radiation Superconducting Shield), 175

Starfish Prime (high-altitude test), 117–19, 120, 123

Star Trek, 1, 14–15, 45, 50, 76–77, 167–68

"Star Wars." *See* SDI (Strategic Defense Initiative)

Star Wars (movie), 45, 73, 75, 76

State Department, 89–90

Stoudt, David, 78

Strategic Capabilities Office. *See* SCO (Strategic Capabilities Office)

Strategic Defense Initiative. *See* SDI (Strategic Defense Initiative)

Stuxnet, 136–38

submarines, nuclear, 21, 24, 25–26

Sun Tzu, 5, 35, 153, 165, 179, 204

Surface Electronic Warfare Improvement Program, 161

swarm attacks, 3, 25, 79–80, 101–2, 186–87, 190

Syria, 19, 142

Tactical Electronic Warfare System. *See* TEWS (Tactical Electronic Warfare System)

Taffola, Tracy, 96

Taiwan, 28

Tamene, Yared, 134

tanks, 40, 171–72

TASS, 171–72

Technical Research Group. *See* TRG (Technical Research Group)

Teller, Edward, 73

Terminal High Altitude Area Defense (THAAD), 6, 125

terrorism, 52, 60, 61, 157–58, 160

testing, nuclear, 117–20

TEWS (Tactical Electronic Warfare System), 163

THAAD. *See* Terminal High Altitude Area Defense (THAAD)

Third U.S. Offset Strategy, 2–3, 25, 53, 55–58, 60, 61, 219

Thomas, Raymond, 142

ThoughtCo., 119

Thucydides Trap, 211–12

Tillerson, Rex W., 91

TNT as nuclear force measurement, 29–30

Toon, Owen Brian, 32

Townes, Charles, 67, 68–69

Treaty on Principles Governing the Activities of States in the Exploration and Use of Outer Space, including the Moon and Other Celestial Bodies. *See* Outer Space Treaty

Treaty on the Prevention of the Placement of Weapons in Outer Space and of the Threat or Use of Force against Outer Space Objects. *See* PPWT treaty

TRG (Technical Research Group), 68, 69, 71

Trident Juncture (war game), 142, 205–6

Trinity (atomic bomb), 116–17

Triple Canopy, 210

Truman, Harry, 49

Trump, Donald: defense strategy of, 57, 60, 61, 127, 158, 210, 217; election win of, 56, 129–31, 134; Havana Syndrome and, 91; *Hill* on, 28; inauguration of, 53

The Trump White House (Kessler), 100

Ukraine, 59, 142, 160, 216

Ulasen, Sergey, 136

United Kingdom, 31, 51, 197

United Nations, 18, 49, 72, 98, 99, 182, 188, 197

United Press International, 138

United States: active protection system of, 172; automation and, 25–26; autonomous weapons of, 183, 188–91; China and, 28, 84–85, 150, 165–66, 211–12, 218, 223–24; Cold War and, 18–19, 214–15; Cuban Missile Crisis and, 119–20; cyberspace abilities of, 21, 128, 137–38, 147, 149–50; defense budget of, 40–41, 56–57, 141, 223–24; directed-energy weapons and, 22–23, 37, 44–45, 105; EMP weapons and, 117, 119, 123–24, 125–26, 127–28, 157–58, 159–60; in Korean War, 49–50; laser weapons of, 42–43, 88, 217; MAD and, 20–21; microwave weapons of, 95; missiles of, 21–22, 24–25, 36; nuclear war effects on, 31–32; nuclear weapons of, 31, 51–52, 104, 119–20, 217–18; pace of warfare and, 34–35; population of, 31; radiation knowledge of, 108–9; rail guns and, 23–24; Russia and, 218; satellites of, 194, 196–97, 200, 207; Soviet Union and, 73, 91; space assets of, 191, 199, 201, 202–6; space weaponization by, 210–11; terrorism distracting, 52–53; threats to, 79, 202–6; treaties and, 20, 22–23, 72, 197, 216–17; weapons buildup of, 4, 15, 22, 144; as world power, 5–6, 19–20, 27, 52, 58–60; World War II and, 48

uranium enrichment, 126, 136, 137

U.S. Air Force, 3, 96, 97, 100–101, 123, 163–65, 179, 193, 207, 210

U.S. Army: countermeasures by, 116, 117, 162–63, 172, 206; laser weapons of, 71–72, 81, 82, 156, 203; microwave weapons of, 97

U.S. Navy: Chinese hacking of, 83; combat system of, 79–81, 143–44; countermeasures by, 161–62; drones and, 184, 186, 219; in First Gulf War, 157; hypersonic missiles and, 24–25; laser weapons of, 4, 36, 75–76, 77, 78, 80–81, 102–3, 155, 156, 186; microwave weapons of, 96, 103; rail guns and, 23; successful missions of, 203; torpedoes of, 45; in war games, 102; weapon system of, 182

U.S. Space Force, 210
USSR (Union of Soviet Socialist Republics).
  See Soviet Union

Van Allen, James A., 118
Van Allen radiation belts, 118
Vangsness, Kirsten, 158
Verge, 25
VirusBlokAda, 136

warfare: asymmetrical, 4, 5–6, 22–23, 60–61,
  141; changing nature of, 1, 8, 42, 77, 165–66,
  179, 190–91, 211–12, 218, 219–22; cyber, 7, 21, 38,
  138, 205–6; electronic, 7–8, 128, 139–44, 147–
  50, 159–60, 161–66, 187, 205–6; speed of, 2–7,
  8, 34–37, 190, 213
war games, 102, 142
War of the Worlds (Wells), 65, 71, 167–68
Warsaw Treaty Organization (Warsaw Pact), 52
Washington, George, 35–36
Washington Examiner, 123–24
Washington Free Beacon, 83
Washington Post, 72, 130
Watson-Watt, Robert Alexander, 140
WaveBand Corporation, 96
weapons, anti-satellite, 21–23, 199–202, 205,
  207–9, *208*, 212
weapons, autonomous: advantages of, 186–87;
  AI and, 2–3, 57; c-war and, 219–21; definition
  of, 180; disadvantages of, 180–82; ethics of,
  182–83, 185–86; in future, 191; major powers
  using, 188–91; malware as, 149; nuclear war
  and, 218–19; overview of, 215–16; pace of warfare and, 8; semiautonomous weapons and,
  183–85; use limitations on, 187–88
weapons, biological, 95–98, 219
weapons, conventional, 22, 34, 37, 44, 59, 72,
  95, 149, 219
weapons, cyber, 7, 21, 135, 137–38, 148–49, 158–
  59, 166
weapons, cyberspace: in American presidential election, 130–32, 133–35; autonomous, 187;
  countermeasures against, 158–61; definition
  of, 135; as directed-energy weapon, 1; ethics
  and, 185; future of, 148–49, 150; overview of,
  128; Stuxnet as, 136–38; tools of, 132–33
weapons, defensive, 4, 75, 170
weapons, directed-energy: advantages of, 43–

44, 95, 186–87; beginnings of, 77; China
  and, 106; countermeasures against, 153–54,
  165–66; c-war and, 219–21; electromagnetic
  energy and, 4; overview of, 1, 215–16; SDI
  and, 73; U.S. military developing, 5, 6, 15, 22–
  23, 37, 44–45, 80, 102–3, 105–6, 161–65. *See
  also* weapons, autonomous; weapons, cyberspace; weapons, EMP; weapons, laser; weapons, microwave
weapons, EMP: countermeasures against, 156–
  58; as directed-energy weapon, 1; Faraday
  cages and, 108; mitigating factors and, 127–
  28; nonnuclear, 121–23; of North Korea, 198;
  nuclear, 116–20, 123–27; result of attacks by, 15–
  18, 31–32, 111–12, 123–27; of U.S. Air Force, 101
weapons, Frey effect, 96–98, 99, 107
weapons, HEMP (high-energy EMP), 156–
  58, 198
weapons, HPM (high-power microwave), 156–
  57, 158
weapons, kinetic, 171
weapons, laser, *45*, *87*; advantages of, 43–44,
  81–82, 186–87; as autonomous weapons, 189;
  capabilities of, 15, 37, 42, 59, 71–72, 194, 203,
  204–5; of China, 82–85; cost of, 6; countermeasures against, 154–56; as defense, 79–
  81; development of, 4–5; as directed-energy
  weapon, 1; ethics and, 72; in future, 190–91,
  215–16; international agreements and, 217;
  microwave weapons compared to, 94–95,
  102–3, 205; of Russia, 82, 85–88; science fiction predicting, 167; technology driving,
  74–75, 76–77; United States developing,
  22, 36–37, 42–43, 71, 73–74, 75–76, 206, 217;
  weather affecting, 77–78, 94
weapons, mass destruction, 22, 123, 199, 213
weapons, microwave: antipersonnel, 95–99;
  attributes of, 94–95; autonomous weapons
  and, 189; capabilities of, 204–5; of China,
  106–7; countermeasures against, 107–9, 155,
  156–58; definition of, 94; as directed-energy
  weapon, 1; and drone defense, 101–3; laser
  weapons compared to, 103, 205; "lights out,"
  100–101; of Russia, 91–92, 104–6; of Soviet
  Union, 90–91; of United States, 103
weapons, nonnuclear, 121–23, 157–58, 220
weapons, nuclear: in arms race, 19–20; of

China, 84; conventional weapons compared with, 59; development of, 51; directed-energy weapons and, 42, 73; EMPs caused by, 116–20; in future, 218–19; HEMP weapons as, 156, 157; human judgment and, 211, 214–15; MAD and, 21–22, 213–18; measurement of, 28–30; in memory only, 27, 28; Milstar and, 195; nations holding, 31, 51–52; of North Korea, 198; radiation caused by, 107; of Russia, 220; of world powers, 27–28, 37, 44, 48, 49–50, 203, 212

weapons, offensive, 79, 187–88

weapons, projected-energy, 1, 76–77, 92, 165–66, 213, 218

weapons, projectile, 4, 23

weapons, semiautonomous, 2, 183–85, 218–21

weapons, strategic, 123–24

weapons, tactical, 123, 163

weapons, thermonuclear, 27, 30, 32, 51, 121, 122

weapon systems, 45, 75–76, 78, 83, 180, 182–83, 186–87, 215–16, 218–19

Weber, Joseph, 66–67

Weinberger, Caspar W., 86

Wells, H. G., 65, 71, 167–68

"white hats" (hackers), 159

white papers, 160

Wilson, Robert, 116–17

Wilson, Tom, 204

Wired, 118, 145

Work, Bob, 3, 55–56

World War I, 35

World War II, 13, 18–19, 27, 28–29, 35, 47–48, 139–41

World War III, 9, 98

X-37B (Orbital Test Vehicle), 207–9, *208*, 210

X-47B (autonomous weapon), 184–85, 189, 219

X-Agent (malware), 132

X-rays, 73, 122, 173–74

Yamamoto, Isoroku, 48

Yanjun Xu, 146

Zeiger, Herbert, 67

Z machines, 122

Zumwalt, USS, 23